Lecture Notes in Artificial Intelligence 1437

Subseries of Lecture Notes in Computer Science
Edited by J. G. Carbonell and J. Siekmann

Lecture Notes in Computer Science

Edited by G. Goos, J. Hartmanis and J. van Leeuwen

Springer
Berlin
Heidelberg
New York
Barcelona
Budapest
Hong Kong
London
Milan
Paris
Singapore
Tokyo

Sahin Albayrak   Francisco J. Garijo (Eds.)

Intelligent Agents
for Telecommunication
Applications

Second International Workshop, IATA'98
Paris, France, July 4-7, 1998
Proceedings

 Springer

Series Editors
Jaime G. Carbonell, Carnegie Mellon University, Pittsburgh, PA, USA
Jörg Siekmann, University of Saarland, Saarbrücken, Germany

Volume Editors

Sahin Albayrak
Technische Universität Berlin, FR 6-7
D-10587 Berlin, Germany
E-mail: sahin@cs.tu-berlin.de

Francisco J. Garijo
Telefonica I+D
C/Emilio Vargas 6, 28043 Madrid, Spain
E-mail: fgarijo@tid.es

Cataloging-in-Publication Data applied for

Die Deutsche Bibliothek - CIP-Einheitsaufnahme

Intelligent agents for telecommunication applications : second
international workshop ; proceedings / IATA '98, Paris, France, July
4 - 7, 1998. Sahin Albayrak ; Francisco J. Garijo (ed.). - Berlin ;
Heidelberg ; New York ; Barcelona ; Budapest ; Hong Kong ;
London ; Milan ; Paris ; Santa Clara ; Singapore ; Tokyo : Springer,
1998
  (Lecture notes in computer science ; Vol. 1437 : Lecture notes in
  artificial intelligence)
  ISBN 3-540-64720-1

CR Subject Classification (1991): C.2, I.2.11, H.4.3, H.5

ISBN 3-540-64720-1 Springer-Verlag Berlin Heidelberg New York

© Springer-Verlag Berlin Heidelberg 1998
Printed in Germany

Typesetting: Camera ready by author
SPIN 10637972   06/3142 – 5 4 3 2 1 0   Printed on acid-free paper

Preface

The first international workshop on Intelligent Agents for Telecommunications Applications (IATA) was held in July 1996 at Budapest during the XII European Conference on Artificial Intelligence ECAI 96. The workshop program consisted of technical presentations, which addressed agent based solutions in areas such as network architecture, network management, and telematic services. The presentations gave rise to a lively debate on the advantages and difficulties associated with incorporating agent technology in telecommunication applications. The proceedings were published by IOS Press providing introductory papers on agent technology as well as telecom applications and services and also papers about appropriate languages and development tools.

Agent technology is a very promising approach to address the challenges of modern day telecommunications. The existing world of telecommunications – which is deeply influenced by monopolistic public network operators (PNOs) – is currently changing at a rapid pace. This change is taking place in the technological as well as the regulatory arena. Additionally, market forces on an unprecedented scale are at work. Given these circumstances it will no longer be sufficient for PNOs to solely provide network infrastructure. The challenge for PNOs consists in evolving to full-service providers. This implies that on the one hand increasingly complex telecommunications infrastructure needs to be managed more efficiently and, on the other hand, that new types of telecommunications services need to be developed and provided. It is in particular such future services that need to satisfy a diverse range of requirements, e.g. personalization, support for user mobility, on-demand combination of different services, offline/online service usage, etc.

Agent technology addresses these requirements particularly well as opposed to other technologies, e.g. client-server. A stationary agent can reside on agent platforms "in the net", providing various types of services. Besides being potentially decentralized and cooperative, these stationary service provider agents possess capabilities for such issues as security, accounting and billing, etc. On the client side, agent-based services will be requested by means of small, mobile agents which may enable both offline and online service usage. Agent technology is very well supported by the language Java and corresponding Java APIs.

IATA'98 will take place in Paris in the framework of Agents' World, which bring together the principal scientific and technical events on agent technology such as the International Conference on Multi-Agent Systems (ICMAS'98), RoboCup'98 devoted to international competition between soccer-playing robot teams, and six international workshops. Each workshop focuses on specific aspects of agent technology like Databases and information discovery in INTERNET (CIA'98), Collective Robotics (CRW'98), Simulation (MABS'98), Agent Theories, Architectures and Languages (ATAL'98), Communityware (ACW'98), and Telecommunications Applications (IATA'98)

The aim of IATA'98 is to provide a state-of-the-art forum for presenting innovative agent based telecommunications applications, and for discussing new approaches, new models and technology trends in both telecom and agent related fields.

This volume contains revised versions of the papers selected by the program committee for presentation and discussion in IATA'98.

The book comprises a collection of seventeen papers organized into five groups. Contributions in the first group present new models of *Network Architecture* using different approaches such as hierarchies of agents, genetic algorithms, and co-operative mobile agents.

The second group deals with new approaches for *Network configuration and planning*. Three models are described. The first model is based on a distributed cognitive agent control architecture, the second model uses a reactive architecture based on ant-like agents, and the third proposes mobile agents.

Two papers in the third group address *Network Optimization* issues and dynamic resource allocation. The first contribution proposes several methods for dynamic resource allocation using the plug and play approach. The second paper describes an architecture based on the interaction of self-interested agents.

The fourth group on *Network Management* contains three papers two of which describe different models of multi-mobile-agent architecture for fault location. The first is based on a cognitive architecture for solving software problems in telecommunication networks, while the second describes a reactive architecture inspired by the foraging activities of ants, where chemical interaction is the principal mechanism for agent communication. The third paper addresses the monitoring and control of network and system components using both stationary and intelligent mobile agents.

The last group contains six papers addressing *Agent based architecture for service applications*. The first paper focuses on the use of agent metaphors for analyzing problematic cases of global inconsistency in distributed information systems. The second describes a multi-agent testbed for seamless messaging and intelligent network management. The next two papers present two different frameworks for negotiating agents. The first is for supporting personal mobility, and the second is for intelligent user interface for cooperative work services. In both cases, the agents negotiate on behalf of the user the best conditions relating to computing and communications resources in terms of quality of service and costs. The fifth paper reports on experimentation and evaluation, of several agent platforms, mixing both the intelligence and the middleware aspects.The last paper describes a Java based agent architecture for building applications concerned with electronic commerce and telecommunication services.

Acknowledgments

We would like to express our sincere gratitude to all the people who helped to bring about the production of this book. The Agents' World chairman Yves Demazeau gave the idea for this second IATA workshop, providing continuous support for both workshop organization and paper publication. His role is gratefully acknowledged.

Nothing would have been possible without the initiative and dedication of the DAI-Lab team at the Technical University of Berlin.

We owe particular gratitude to the members of the program committee for their professionalism and dedication in selecting the best papers for the workshop. We especially thank all contributing authors for choosing IATA'98 to present their research results, and for their diligence and their cooperation in the preparation of this volume.

Hans Schlenker of the DAI-Lab has organized the review process, keeping in touch with the authors and monitoring the submitted contributions and the accepted papers. He did a great job.

We would like to thank Telefónica I+D for providing the environment and the technical facilities to prepare the book. Thanks also to Susana Suarez who handled the book formatting according to Springer guidelines, and Jose Maria Matias for his help in solving word processing and text editing troubles.

Finally, we would like to express our appreciation of the workshop sponsors:

- Deutsche Telekom
- France Telecom
- Sun Microsystems
- Alcatel
- Siemens AG

Paris, July 1998

Sahin Albayrak	Francisco J. Garijo	Michel Plu
Workshop Chair	Publication Chair	Program Chair

Organizing committee

Chairman
Sahin Albayrak (Chair)
Technische Universität Berlin Berlin Germany
Co Chairmans
Francisco J. Garijo
Telefónica I+D Madrid , Spain
Charles Petrie
Center for Design Research, Stanford University, Stanford, CA USA
Michel Plu
France Telecom, CNET, Lannion , France

IATA´98 Program committee

Frank von Martial, Germany
Fumio Hattori, Japan
Danny Lange, USA
Divine Ndumu, UK
Hyacinth Nwana, UK
Innes Ferguson, Canada
Katia Sycara, USA
Leonardo Chiariglione, IT
Mark Fox, Canada
Mike Wooldridge, UK
Munindar P. Singh, USA
Nick Jennings, UK
Paul Kearney, UK
Peter Selfridge, USA
Robert Weihmayer, USA
Tim Finin, USA
Toru Ishida, JAPAN
Victor Lesser, USA   (invited)
Walter Van de Velde, Belgium
Yves Demazeau, FR

Table of Contents

A Dynamic Hierarchy of Intelligent Agents for Network Management

Christian Frei and Boi Faltings

Artificial Intelligence Laboratory
Swiss Federal Institute of Technology (EPFL)
IN-Ecublens, CH-1015 Lausanne, Switzerland
{frei, faltings}@lia.di.epfl.ch

Abstract. Routing as well as the management of communication networks that support hybrid types of communications requiring quality of service is a hard problem. We present here a framework[1] that decomposes the network into a hierarchy of abstract views of the network that summarizes the available bandwidth resources and highlights bottlenecks in the network in order to reduce the complexity of the previously mentioned tasks. This framework can easily be distributed to a hierarchy of intelligent agents. This framework is technology independent and can be applied to any connection-oriented communication network.

1 Introduction

Many applications, especially those related to multimedia, require that information be transmitted in a communication network with certain quality of service guarantees. Such guarantees can only be given in connection-oriented networks, such as ATM [7]. However, if the routing can be solved efficiently by local algorithms in packet-switched networks, this is not the case for connection-oriented networks. In a connection-oriented network, an information flow cannot be split between several different routes: one and only one route can be allocated for a transmission, and resources must be reserved along the chosen route.

A communication network is a set of nodes that are interconnected by links to permit the exchange of *information*. We model it as a network graph (Fig. 1). A need to exchange information between two nodes is called a *demand*. Each demand requires a certain *quality of service* (QoS) when transferred through the network. The required QoS depends on the type of information to transmit and is constrained using several parameters, such as minimal bandwidth, maximal delay, maximal delay jitter, or maximal cell loss rate. In order to satisfy a demand, we must allocate a *route*, between the two endpoints of the demand, respecting the demand's QoS constraints. Assigning a route to a demand and reserving the resources needed is called

---

[1] A patent on the work presented in this paper is pending.

establishing a connection. Given a communication network, the problem is to allocate one route for each incoming demand in the network so that the QoS constraints of all demands are satisfied using the available resources of the network. This problem is known to be NP-complete [14].

We propose here a general framework using abstraction techniques [1,5] called *Blocking Islands*. The idea is to build an *abstraction* of the original communication network into a hierarchy of simplified graphs where each node, a blocking island, clusters a part of the network inside which routing of demands requiring a given amount of bandwidth is possible. Each link of an abstract graph then identifies bottleneck links of the network. This hierarchy can be adapted dynamically in polynomial time to reflect the changes in the network's state, such as allocation or deallocation of connections. The information contained in the hierarchy can be purposefully used to derive efficient heuristics that achieve a load balancing effect on the network's links. The management of this hierarchy as well as the routing of demands can easily be distributed to intelligent agents.

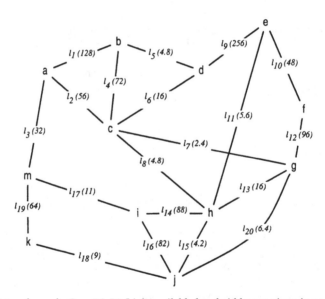

Fig. 1. Network graph $G = (N, L)$. Link available bandwidths are given in parenthesis.

2   Problem modeling

A communication network is composed of nodes interconnected with communication links. We model it as a connected *network graph* $G = (N, L)$ (Fig. 1), a non-directed multi-graph without loops[2], where the nodes N are processing units, switches, etc.,

---

[2] A loop in a multi-graph is a link whose endpoints are the same node.

and the links L correspond to bi-directional communication media, such as optical fibers. Each link l is characterized by its *bandwidth* resources β_l, typically measured in *[bits/second]*, and other properties such as delay (*[ms]*) and cell loss rate.

Our network must fulfill communication needs between pairs of nodes, or *demands* (e.g., phone calls, video conferencing, video on demand, etc.). A demand d_k is a triple:

$$d_k = (x_k, y_k, \beta_k) \tag{1}$$

where x_k and y_k are nodes of the network graph G and define the nodes between which communication is required to take place: the demand's *endpoints*. The following parameter describes a *quality of service (QoS)* requirement: β_k is the amount of *bandwidth* needed, or information exchange rate. In this paper, we suppose that β_k is a constant number (*constant bit rate* connections - CBR). Other QoS requirements (such as delay, delay jitter, maximum cell loss rate) may also characterize a demand and Sect. 6 shows how to take them into account.

In order to satisfy a demand, we must allocate a *route*, between the two endpoints of the demand, respecting the demand's QoS requirements in order to define an *information transmission media*. A route is a simple path[3] of the network graph and thus a set of links. A *routing algorithm* ensures route computation. Assigning a route to a demand and reserving the resources needed is called *establishing a connection*.

When establishing a connection for a demand d_u, bandwidth resources are reserved among the links that build the route in order to ensure QoS. Verifying that a link actually has enough available bandwidth to support d_u and the amount of resources that must be reserved depends on the underlying network technology and is driven by the *Connection Admission Control* (CAC) of each link. In the following, we suppose a Boolean function $\alpha(l, \Gamma, \beta_u)$ that returns TRUE if and only if link l has enough available bandwidth for the unallocated demand $d_u=(x, y, \beta_u)$ given the already established connections Γ. α is obviously network technology dependent and takes any overhead or overbooking techniques into account. A route r has then enough available bandwidth for d_u given the established connections Γ if and only if all links of r have enough available bandwidth for d_u.

In order to ease the understanding of the notions introduced in this paper, we define that the *available bandwidth* on a link is the maximum amount of bandwidth that can be allocated to a demand given the already established connections. In other words, a link with k available bandwidth can support a demand requiring at most k bandwidth resources: $k = \max \{\beta : \alpha(l, \Gamma, \beta) = \text{TRUE}\}$.

Our problem is to find *one and only one* route (recall that we only consider connection-oriented networks) for each demand satisfying the bandwidth constraints of the demands using the available resources of the network.

In general, demands do not arise all simultaneously in a communication network. Our problem is thus best defined as an incremental allocation where demands must be routed as they arrive. Allocating routes for the first demands has an incidence on the routing of the following demands since the former use up some resources that are

---

[3] A *simple path* is a path where no subpath is a cycle.

therefore not available anymore to the latter. The first established connections may then cause the routing of an additional demand impossible. The number and the QoS requirements of the future incoming demands are generally unknown and therefore there is no way to optimize the allocation in order to ensure that all demands can be routed.

3 Related Work

Most applications of Artificial Intelligence in communication networks have focused on the problems of fault diagnosis and fault management. The most recent overview of applications of knowledge systems in telecommunications networks can be found in [11]. Resource allocation has been recognized as a problem more recently. The importance of resource allocation for network performance, resource utilization and its impact on the profitability (of the service provider), and the significance for the network users is discussed in [12]. Franken and Haverkort [8] have worked on a performance manager, which proposes network reconfigurations to improve performance. [10] discusses an expert system called ANMA under development for ATM networks. It addresses fault diagnosis and management, but also includes modules for collecting performance information and suggesting changes to routing tables and network topology to improve performance. ANMA is the only tool which implements a connection between network planning and routing decisions, but it does so in a very rudimentary and ad-hoc fashion.

More lately, a lot of research effort has been devoted to QoS routing, such as in [16], [15] and [17]. The algorithms they propose are heuristics-based in order to allow real-time computation, since this problem is NP-complete [14]. [2] presents methods to allocate routes on-line to virtual circuits. [4, 6] propose methods for route selection using pre-computation in ATM networks.

A key research in communication networks nowadays is how to move management from centralized to distributed, in order to overcome the increasing complexity of this task, due to network size and emerging technologies, such as ATM, that compound this problem by requiring that large quantities of performance data be processed and analyzed for maintaining performance guarantees. Most solutions, such as *P-NNI* [3] and *HYBRID* [13], are based on a hierarchy of autonomous intelligent agents that have local decision-making capabilities, but co-operate to resolve conflicts. Higher level agents arbitrate unresolvable disputes between peer agents.

In these systems, an agent reigns over an arbitrarily and statically defined subnetwork, in most cases an administrative domain of some kind. However, these domains do not reflect the easy routing parts of the network and do not evolve: heuristics for efficient routing by management are therefore difficult to apply in these cases.

4   Blocking Islands

The Blocking Island paradigm has been introduced in [9]. We recall here its most important results.

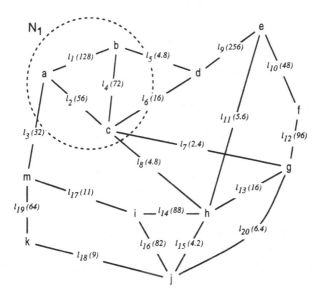

Fig. 2. Network graph $G = (N, L)$ where the weights on the links refer to their available bandwidths. $N_1 = \{a,b,c\}$ is the $64K$-blocking island ($64K$-BI) for a.

4.1   β-Blocking Islands

After a set of demands has been allocated, we call a β-blocking island for a node x, the set of all nodes of the network that can be reached from x using links with at least β available bandwidth:

Definition 1 (β-Blocking Island)   *Given a network graph $G = (N, L)$, a set of established connections Γ, a node $x \in N$ and a bandwidth requirement β, the set of nodes $S \subseteq N$ is a β-Blocking Island (β-BI) for x if:*
   1. $x \in S$
   2. $\forall y \in S - \{x\}$, $\exists$ route r_i between x and y, $\alpha(r_i, \Gamma, \beta) = TRUE$
   3. $\forall l \in \omega(S)$, $\alpha(l, \Gamma, \beta) = FALSE$
where $\omega(S)$ is the cocycle[4] of S. The links of $\omega(S)$ are called β-Blocking Island Links (β-BILs).

---

[4] The *cocycle* of a subset of nodes A is the set of all links that have one and only one endpoint in A.

Fig. 2 shows the *64K*-BI for node *a*. Note that it is false to say that all links inside a β-BI, i.e., the links that have both endpoints in the β-BI, have at least β available bandwidth: there may be a link *l* with less than β available resources inside a β-BI. In such a case, it simply means that there is another route with β available bandwidth between *l*'s endpoints. As a matter of fact, in Fig. 2, link l_2 has both endpoints in the *64K*-BI but has less than 64 available resources. However, there is at least 64 available bandwidth on route $\{l_1, l_4\}$ between l_2's endpoints.

β-BIs have some fundamental properties (the proofs of these are given in [9]) that illustrate their usefulness in routing decidability:

- **Unicity:** There is one and only one β-BI for a node, $\forall\beta$. Thus, if *S* is the β-BI for a node *x*, *S* is the β-BI for every node in *S*.
- **Partitioning:** β-BIs induce a partition of a network's nodes since they define equivalence classes over the nodes.
- **Bottleneck identification:** The links of a β-BI's cocycle, the β-BILs, do not have enough available resources to support a demand requiring β (or more) bandwidth.
- **Route existence:** There is a route satisfying the bandwidth requirement of an unallocated demand $d_u=(x, y, \beta_u)$ if and only if its endpoints *x* and *y* are in the same β_u-BI.
- **Route location:** The links of a route with β available bandwidth are all in the β-BI of its endpoints.
- **Inclusion:** If $\beta_i < \beta_j$, the β_j-BI for a node is a subset of the β_i-BI for the same node.

The β-BI *S* for a given node *x* of a network graph can be obtained with a simple greedy algorithm, with a linear complexity of $O(m)$, where *m* is the number of links in the network [9].

4.2   β-Blocking Island Graphs

We use the properties of BIs (Sect. 4.1) to build a β-blocking island graph, a simple graph representing an *abstract* view of the available resources: each β-BI is clustered into a single node and there is a abstract link between two of these nodes if there is a β-BIL joining them, i.e., if their cocycle has a common β-BIL. These abstract links denote the critical links, since their available bandwidth is not enough to support a demand requiring β resources.

Definition 2 (β-Blocking Island Graph). *The* β*-Blocking Island Graph (*β*-BIG) for a network graph G = (N, L) given a set of established connections* Γ *is an abstract simple graph* $G_I = (N_I, L_I)$ *of G where:*

- N_I *is the set of all* β*-blocking islands of G given* Γ*. Such a node is called* abstract node, β*-node or just* β*-blocking island.*
- *There is a link* $L \in L_I$ *between two nodes* S_i *and* S_j *of* N_I *if and only if there is a least a link of L that has one endpoint in* S_i *and the other in* S_j*: L is then the set*

of all such links of L, thus $L = \omega(S_i) \cap \omega(S_j)$. *A link of* L_I *is called an* abstract link *or a* β-link.

- *The* available bandwidth *on an abstract link is the maximum of the available bandwidth on its children links*[5].

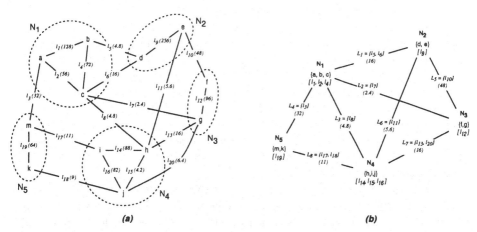

Fig. 3. *64K*-Blocking Island Graph: (a) shows a network graph where the links' weights (in parenthesis) denote their available bandwidth; (b) is the corresponding *64K*-BIG, where the abstract links' weights (in parenthesis) are their available bandwidth. Children of an abstract link are given in brackets, whereas children of an abstract node are split into two sets: the node children (in brackets) and the link children (in square brackets).

The route existence property (Sect. 4.1) is easily extended to a β-BIG: there is a route satisfying the bandwidth requirements of $d_u = (x, y, \beta_u)$ if and only if the endpoints of d_u, x and y, belong to the same β_u-node in the β_u-blocking island graph.

The abstraction power of a β-BIG is not limited to clustering network nodes, since network links that have both endpoints in a β-BI are clustered in that β-BI in the BIG: only critical links, clustered into abstract links, are represented in it. For instance, in Fig. 3, l_5 is a *64K*-BIL for N_1 and N_2 and is clustered in abstract link L_1, whereas l_1 does not appear in any abstract link of the *64K*-BIG, because it is clustered in β-node N_1.

A β-BIG is an abstraction of a network graph; these abstractions are described by father-child relations:

- The *father* of a network element, i.e., a node or a link, is the β-BIG element (a β-node or a β-link) that clusters it.
- The *children* of a β-BIG element are the network elements it clusters. In order to distinguish the two types of children a β-node can cluster, the children are split into *node children* and *link children*.

The construction of a β-BIG is straightforward form Definition 2 and is polynomial in $O(m)$ where m is the number of communication links [9].

---

[5] Since a demand can only be allocated one route.

5 Blocking Island Hierarchy

A β-blocking island graph presents only the bottleneck links for a given β. These links cannot support a demand requiring β or more bandwidth. It would be however useful to know the bottlenecks for different βs, e.g., for typical possible bandwidth requirements. A possible but costly solution is to compute a BIG for each β. This does however not take advantage of the inclusion property of BIs. A better alternative is thus to build a recursive decomposition of blocking island graphs in decreasing order of the βs: $\beta_1 > \beta_2 > ... > \beta_b$. We call this layered structure of BIGs a *Blocking Island Hierarchy* (BIH). The lowest level of the blocking island hierarchy is the β_1-BIG of the network graph. The second layer is then the β_2-BIG of the first level, i.e., β_1-BIG, the third layer the β_3-BIG of the second, and so on. On top of the hierarchy we have a *0*-BIG abstracting the smallest bandwidth requirement β_b. The abstract graph of this top layer is reduced to a single abstract node (the *0*-BI), since the network graph is supposed connected.

Fig. 4 shows such a BIH for bandwidth requirements *{64K, 19.2K}*. The graphical representation shows that each BIG is an abstraction of the BIG at the level just below (the next biggest bandwidth requirements), and therefore for all lower layers (all larger bandwidth requirements).

A BIH can not only be viewed as a layered structure of β-BIGs, but also as an *abstraction tree* when considering the father-child relations. The abstraction tree of the BIH of Fig. 4 is given in Fig. 5. In an abstraction tree, the leaves are network elements (nodes and links); the intermediate vertices either abstract nodes or abstract links and the root vertex the *0*-blocking island of the top level in the corresponding BIH. We extend the notions of father and child (Sect. 4.2) as follows:

- The *network children* of an abstract node or link are the network elements it clusters. The network children of an abstract node are divided into *network node children* and *network link children*. The network children of an element are the leaves of the subtree whose root is that element in the abstraction tree. For instance, in Fig. 5, the network children of *19.2K*-node N_6 are *{a,b,c,k,m}* and its network link children *{l_1, l_2, l_3, l_9, l_{10}}*. Similarly, the network children of L_9 are *{l_5, l_6, l_7}*.
- Conversely, the β-*father* of a network node or link is the β-node or β-link that abstracts the network element at β level in the BIH. The β-father of a network element is obtained by following the father links in the abstraction tree until level β is reached. For instance, in Fig. 5, the *19.2K*-father of network node k is N_6.

A BIH for a set of constant bandwidth requirements ordered decreasingly is easily obtained by recursive calls to the BIG computation algorithm. Its complexity is bound by O(b m), where m are the number of communication links and b the number of bandwidth requirements.

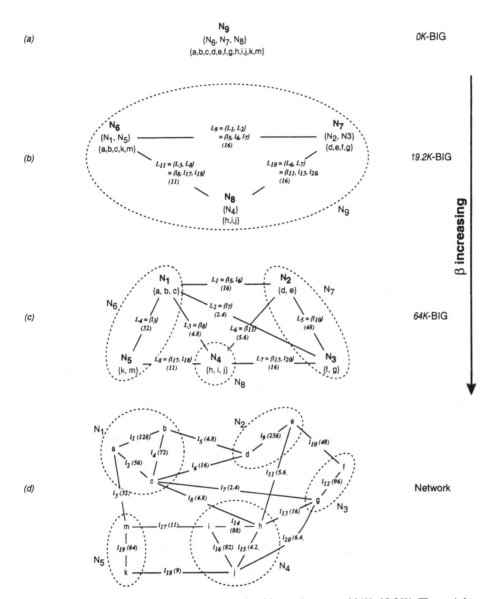

Fig. 4. The blocking island hierarchy for bandwidth requirements *{64K, 19.2K}*. The weights (in parenthesis) on the links are the available bandwidths. Abstract links' description includes their children and network children in brackets. Abstract nodes' description includes only their node children and network node children in brackets: link children are omitted for more clarity. (a) the *0*-BIG; (b) the *19.2K*-BIG; (c) the *64K*-BIG; (d) the network graph.

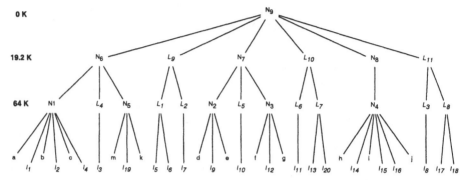

Fig. 5. The abstraction tree induced by the father-child relations on the nodes and links of the blocking island hierarchy of Fig. 4.

6   Maintaining the Blocking Island Hierarchy

The blocking island hierarchy summarizes the available bandwidth given the currently established connections at a time t. As connections are allocated or deallocated, available bandwidth changes on the communication links and the BIH may need to be modified to reflect this. The changes can be carried out incrementally, only affecting the blocking islands which participate in the demand which is being allocated or deallocated:

- When a new demand is allocated along a particular route, the bandwidth of each link l decreases. If it falls below the bandwidth β of its blocking island, and no alternative route exists with capacity $\geq \beta$ within the BI, it causes a split of the BI into two parts. Furthermore, this split must be propagated to all BI in the hierarchy with a higher β. Fig. 6 illustrates the splittings required in the BIH of Fig. 4 after allocating a new connection.

- When a demand is deallocated, bandwidth across each link increases. If it thus becomes higher than the β of the next higher level in the hierarchy, it will cause two disjoint blocking islands to merge into a single one. This merge is again propagated to all levels with a lower β. Fig. 7 illustrates the merges that follow the deallocation of a connection in the BIH of Fig. 4.

[9] provides polynomial-time algorithms that allow the incremental adaptation of a BIH, not only when new connections are established or existing ones deallocated, but also in case of link failure, link properties alteration or even network topology changes.

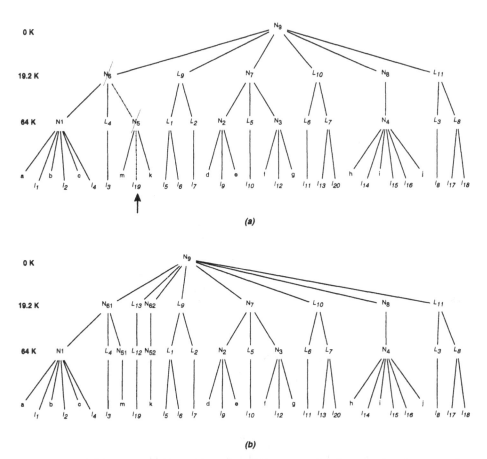

(a)

(b)

Fig. 6. Updating the BIH of Fig. 4 when allocating a connection between k and m, reserving thereby 54K additional resources on link l_{19}. l_{19} has then 10K available bandwidth: lowest splitting point is therefore at β=64K. (a) is the original abstraction tree, where stricken out nodes have to be split: at level 64K, N_5 is split into N_{51} and N_{52}, whereas N_6 is split into N_{61} and N_{62} for the *19.2K*-BIG. (b) is the updated abstraction hierarchy.

7 Routing using the Blocking Island Hierarchy

The route existence property (Sect. 4.1) states that there is a route with β available bandwidth between any two nodes of a β-blocking island. Therefore we know there is at least one route r satisfying a demand $d_u=(x, y, \beta_u)$ if the β_u-father N of x is also the β_u-father of y in the blocking island hierarchy. Moreover, r is inside the β_u-father (route location property – Sect. 4.1), i.e., the links of r have all both endpoints in N. Whatever routing algorithm is used, search space is reduced if we confine it to the subgraph abstracted by N.

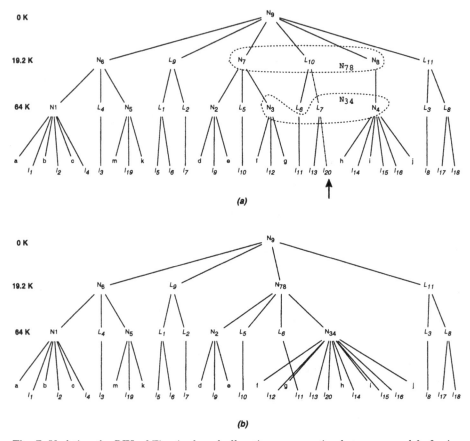

Fig. 7. Updating the BIH of Fig. 4 when deallocating a connection between g and h, freeing thereby 60K additional resources on link l_{20}. l_{20} has then 66.4K available bandwidth: lowest merging point is therefore at $\beta=64K$. (a) is the original abstraction tree, where abstract nodes to be merged at 64K and 19.2K are surrounded. Resulting abstract nodes are N_{34} and N_{78}, respectively. (b) is the updated abstraction tree.

Even better, the search space of the routing algorithm should be set to the β_i-BI ($\beta_i \geq \beta_u$) abstracting both endpoints at the lowest level in the BIH in order to restrict it even more. We call this the *lowest level heuristic*. Besides, this heuristic has two beneficial side effects: the more search space is reduced, the fewer alternative routes are available, therefore facilitating the choice of a route; it also achieves a load balancing effect because the lower N is in the BIH, the more available bandwidth are on the links it clusters.

A BIH gives also the means to compare *a priori* equivalent routes in order to decide for the "best" one, besides the length criterion (the shorter the route, the less resources are used). We know that the higher a link is abstracted in the BIH, the more bandwidth critical it is. Comparison of two routes can then be based on the criticalness of its links, or at which level in the BIH each route is "inside" a BI.

Another way to compare routes is to analyze the consequences each route has on the BIH. The best route is the route that causes the less splittings of abstract nodes in the BIH: obviously, the more splittings, the more links become critical leading to more allocation failures of demands because no route satisfying the new demands are available. This "forward-checking" like criterion is called the *minimum splitting heuristic*, and has also a load balancing effect.

When routing a demand is not only subjected to bandwidth requirement but also to other QoS parameters, such as delay and loss rate, a BIH acts as a first filter: we know if there is a route satisfying the bandwidth requirement and in which part of the network graph to look for it. A route satisfying the other QoS constraints can then be searched in that subgraph. Hereby, as possibly much time is gained through search space reduction, more complex routing algorithms can be used then before (for instance, with constraints on more QoS parameters), or more alternative routes can be examined in order to choose the best one. As many end-to-end QoS parameters are not independent from the available bandwidth resources in the network, e.g., if the load increases, delay, delay jitter and loss increase too, using a bandwidth filter such as a BIH proves to be a very good choice, besides the fact that bandwidth is a critical QoS parameter for any multi-media transmission. This explains why existing QoS routing algorithms [16, 15, 4] always take bandwidth requirements into account and may ignore other parameters.

In short, the benefits of the BIH for routing are twofold:

- Straight answer about whether a route exists satisfying the bandwidth requirement of a demand.
- Problem size reduction, possibly drastically.

8 Distributed Resource Allocation by Agents

Most distributed network management systems, such as *P-NNI* [3] and *HYBRID* [13], are based on a hierarchy of autonomous intelligent agents that have local decision-making capabilities, but co-operate to resolve conflicts. Higher level agents arbitrate unresolvable disputes between peer agents. In these systems, an agent reigns over an arbitrarily and statically defined subnetwork, in most cases an administrative domain of some kind. For instance, a university could be managed as follows: an agent is in charge in each lab, whereas other agents each oversee a department and a single one rules the university. These domains do not reflect the easy routing parts of the network and do not evolve. Therefore, the agents do not have all relevant and important information to take good decisions, leading to substantial negotiations between agents that can result in strong latency of connection establishment.

However, a blocking island is a summarization of resource availability between pairs of nodes and heuristics for efficient routing can be derived from this information (Sect. 6). We propose to use a similar hierarchical architecture of agents as in *P-NNI* or *HYBRID*, to the difference that domains correspond to the blocking islands, and are therefore dynamic.

14

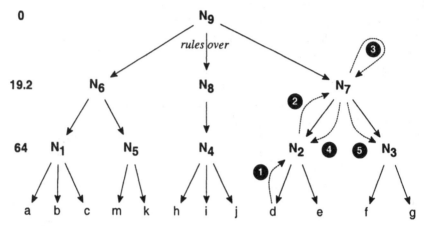

Fig. 8. Distributed resource allocation when d wants to communicate with g at 19.2 Kbits/sec. (1) d asks its direct leader, the agent responsible for N_2, to establish a connection to g with 19.2 Kbits/sec. (2) since N_2 is responsible for 64K connections, it passes the request to its own leader N_7. (3) N_7 is at the correct level and therefore is responsible for connection establishment. Since the other endpoint, g, is one of its network children (via N_3), there is at least one route satisfying the new demand. N_7 first finds a route in its domain connecting N_2 and N_3; there is only one here: $\{l_{10}\}$, abstracted by L_5 (Fig. 4). This route connects e to f. N_7 then delegates the responsibility to complete the route to its child agents N_2 and N_3. (4) N_2 completes the route by looking for one in its domain between d and e: there is only one possibility, over link l_9. (5) N_3 completes its part of the task by finding a route between f and g: $\{l_{12}\}$. Hereby the route $\{l_9, l_{10}, l_{12}\}$ was found and the required resources reserved: communication can take place.

The management and maintenance of a BIH is easily distributed to intelligent agents. Each blocking island is ruled by an autonomous agent, a *BI leader*, which is responsible for resource allocation and management in its domain. Each agent is ruled by one and only one father on the next BIH level. The top-level leader is then the highest agent in charge. Our hierarchy of agents yields the abstraction tree of the BIH and can be managed the same way as the hierarchical systems described above, except that a BIH is dynamic, with same performance gain. Our agents are therefore dynamic and have additional responsibility: the creation and destruction of agents when child blocking islands are split and merged, respectively.

Our agents effectively manage connection establishing. When a new demand arises, issued by a network node x which needs to communicate with another node y at β bits/sec, x asks its direct leader (its father) to establish that connection. The leader passes on the demand characteristics to its own leader, until the leader at β level is reached. The latter is then responsible for establishing a connection for the new demand, that is, find a route satisfying its QoS constraints and reserve the required resources among the chosen route. The β-leader does not have to communicate with peer agents in order to negotiate resources, because any route satisfying the new demand only uses resources he is responsible for (route location property of a BI – Sect. 4.1). The β-leader may however communicate with subordinate agents when finding a route. Fig. 8 gives an example of distributed resource allocation.

The construction of the BIH (at network boot) can also be distributed: a first agent, the top leader, is responsible for computing the BIG for the smallest bandwidth requirement. It then creates a child agent for each BI found and delegates them the duty to partition their domain according to the next smallest bandwidth requirement. This is done recursively until agents for the biggest bandwidth requirement are established.

The advantage of our distributed BIH over systems such as P-NNI and HYBRID is that our domains reflect current bandwidth resource availability: an agent leading a β-BI can allocate β bandwidth between any two network nodes of the β-BI. Routing, and resource reservation are hereby purposefully distributed according to the scheme presented in Sect. 6. As a result, less co-operation and negotiation between BI leaders are required than between agents of the other systems when connections must be established: our agents are more autonomous.

A distributed BIH manages bandwidth resources much more efficiently than other distributed systems, while achieving other tasks (such as network monitoring and performance analysis) just as well at the cost of managing dynamic (and possibly mobile) agents. However, we feel that this disadvantage will be outweighed by the advantages outlined above.

9 Conclusion

By distinguishing the "easy" parts from the critical ones in the network, this BIH framework allows a straight answer about whether a route exists satisfying the bandwidth requirement of a demand and the use of existing routing algorithms more efficiently because the problem size is reduced to a blocking island. Network utilization is improved as this decomposition provides to the routing algorithms more knowledge about the network's state than previously available: fewer demands will have to be rejected.

A distributed BIH manages bandwidth resources much more efficiently than other distributed systems, resolving their arbitrary domain weakness, while achieving other tasks (such as network monitoring and performance analysis) just as well.

10 References

1. Dean Allemang and Beat Liver: How can we communicate with computers? Abstractions : their purpose and application in telecommunications, *ComTec : Technische Mitteilungen Swiss Telecom PTT*, 10:948-956, 1995.
2. J. Aspnes, Y. Azar, A. Fiat, S. Plotkin, and O. Waarts: On-Line Routing of Virtual Circuits with Application to Load Balancing and Machine Scheduling, *Journal of the ACM*, 44(3):486-504, May 1997.
3. The ATM Forum: *P-NNI 1.0 Specification*, May 1996.

4. Jean-Yves Le Boudec and Tony Przygienda: A Route Pre-Computation Algorithm for Integrated Services Network, Technical Report TR-95/113, LRC-DI, EPFL, Lausanne, Switzerland, 1995.
5. Berthe Y. Choueiry and Dean Allemang: Abstraction Methods for Resource Management in a Distributed Information Network, *Workshop on Artificial Intelligence in Distributed Information Networks, IJCAI'95*, pp. 99-102, Montréal, Québec, 1995.
6. Olivier Crochat, Jean-Yves Le Boudec, and Tony Przygienda: A Path Selection Method in ATM using Pre-Computation, Technical Report TR-95/128, LRC-DI, EPFL, Lausanne, Switzerland, 1995.
7. Martin de Prycker: *Asynchronous Transfer Mode: Solution for Broadband ISDN*, Prentice Hall, 3rd edition, 1995.
8. Leonard J.N. Franken and Boudewijn Haverkort: The Performability Manager, *IEEE Network*, pp. 24-32, 1994.
9. Christian Frei and Boi Faltings: Simplifying Network Management using Blocking Island Abstractions, Internal Note from the IMMuNe Project, April 1997. Available at http://lrcwww.epfl.ch/immune/index.html.
10. W. Fuller and S. Miksell: Network management using real-time expert systems, *Worldwide Intelligent Systems*, IOS Press, 1995.
11. J. Liebowitz and D.P. (eds.): Worldwide intelligent systems, *IOS Press*, 1993.
12. R. Smith, N. Azarmi, and I. Crabtree: Special issue on advanced information processing techniques for resource scheduling and planning, *British Telecom Technology Journal*, 1995.
13. Fergal Somers: HYBRID: Intelligent Agents for Distributed ATM Network Management, *Working notes of the Workshop on Intelligent Agents for Telecom Applications*, ECAI'96, Budapest, Hungary, 1996.
14. Sundararajan Vedantham and S. S. Iyengar: Bandwidth Allocation Problem in ATM Network Model is NP-Complete, *Information Processing Letters*, ISSN: 0020-0190, pp. 179-182, 1998.
15. Ronny Vogel, Ralf Guido Herrtwich, Winfried Kalfa, Harmut Wittig, and Lars C. Wolf: QoS-Based Routing of Multimedia Streams in Computer Networks, *IEEE Journal on Selected Areas in Communications*, 14(7):1235-1244, September 1996.
16. Zheng Wang and Jon Crowcroft: Quality-of-Service Routing for Supporting Multimedia Applications, *IEEE Journal on Selected Areas in Communications*, 14(7):1228-1235, September 1996.
17. N. Yeadon, F. García, D. Hutchison, and D. Shepherd: Filters: QoS Support Mechanisms for Multipeer Communications, *IEEE Journal on Selected Areas in Communications*, 14(7):1245-1262, September 1996.

Genetic Algorithms as Prototyping Tools for Multi-Agent Systems : Application to the Antenna Parameter Setting Problem

T. LISSAJOUX   V. HILAIRE   A. KOUKAM   A. CAMINADA*

IPSé - LaRIS - rue du Château - 90010 Belfort Cedex - France
* CNET Belfort - 90010 Belfort Cedex - France
thomas.lissajoux@utbm.fr

Abstract. This paper deals with using an evolutionary algorithm (EA) as a prototyping tool to develop and refine a multi agents system (MAS) for problem solving. In the case of (distributed) solving, MAS may lack some knowledge about the solving mechanism. Using a GA as a prototyping tool thus enables to extract heuristics for use in the MAS design. This approach, based upon a model conciliating both EA and MAS perspectives, is tested on the antenna parameter setting problem (APSP) from the field of radiomobile networks. We demonstrate the feasibility and interest of such an approach for complex problems. Moreover, we advocate the use of MAS techniques for the field of radiomobile networks.

1   Introduction

This paper describes the joint use of evolutionary algorithms and multi-agent systems to solve optimization problems. Evolutionary Algorithms (EA) are algorithmic techniques used in optimization and based upon an analogy with natural evolution of biological organisms [7][1]. Multi-Agent Systems (MAS) [5][12] correspond to a distributed view of AI, in which complex systems are thought as a set of interacting autonomous agents.

The approach described in this paper is based upon the use of an EA as a prototyping tool, to develop and refine a MAS. It is tested on the antenna parameter setting problem (APSP) from the field of radiomobile networks, not only in order to solve the problem itself in an optimization perspective, but also to acquire some knowledge about it. It also investigates the application of MAS techniques to the domain of radiomobile networks.

The reasons justifying this use of MAS in this field are twofold. On the one hand, the field of radiomobile networks, as it deals with independent entities such as antennae and mobile receivers, naturally lends itself to a distributed view [15]. On the other hand, solutions found by MAS present some specific features interesting for network applications. It particularly enables one to obtain flexible solutions [9], relatively to both constraints variations and problem specification variations.

However, whereas MAS are usually seen as tools for experimentation and simulation [13], it appears that, for complex problems such as antenna parameter setting, building an effective MAS can prove to be quite difficult. This is due to the lack of sufficient knowledge about the problem. Indeed, MAS, as they correspond to a natural view, do not require a deep problem modeling. But, in the (distributed) problem solving domain, they need some knowledge about the solving mechanism [4].

In the contrary, EA require a deep modeling but, by their very nature, do not require any knowledge about the solving mechanism [7]. The use of a genetic algorithm can be characterized by its relative easiness of implementation and short development time permitted by efficient reuse tools such as genetic libraries [14]. Using a GA as a prototyping tool thus enables us to focus on the problem in order to learn about the exact nature of solutions and extract heuristics.

In order to develop at the same time an EA, whose problem modeling is naturally centralized, and a MAS, corresponding to a distributed view, the model of the problem upon which the two approaches are based has to lend itself to both approaches. It has therefore to conciliate a distributed view and also a global view by expressing the latter as a sum of the local views.

This paper first presents the problem of antenna parameter setting, as well as the constraints and goals considered. It then describes the two approaches, GA and MAS, giving the approach modeling and behavior. It eventually concludes about the joint use of these two techniques for problem solving and discuss related and future work.

2 Antenna parameter setting problem.

We describe here the context of antenna parameter setting and propose a model. We then define the framework of our optimization problem, defining constraints and stating our goals.

2.1 Overview.

The antenna parameter setting problem (APSP) fits into the global process of radiomobile network design [10]. This process is composed of the three following main stages :

- positioning.
- parameter setting.
- frequency allocation.

These stages successively consist in the positioning (locating) of antennae onto the available sites, the setting of their parameters and eventually frequency allocation. The issues related to the whole process deal with both quantitative aspects :

optimization of the network coverage and minimization of the interference between antennae ; and qualitative ones, as detailed later on. The problem of parameter setting, which concerns us here, deals with the above issues : maximizing the area covered by antennae as well as restricting the interference and allowing a given service quality.

The global problem is to determine the optimum setting of the parameters associated with the antennae in charge of the radiomobile coverage of a given area. The antennae are distributed upon the surface to cover, which is modeled as meshes whose size may vary (25 to 100 meters, depending on the type of land).

2.2 Model.

For each mesh m, a propagation model enables, given the altitude and the type of ground (woods, building...) to predict the local variation $Fade_m^{a_i}$ of the radioelectric field emitted by every antenna a_i. The main parameters of an antenna are $Power^{a_i}$, $Tilt^{a_i}$, $Azim^{a_i}$, its emission power, tilt and azimuth. However, we just consider here omnidirectional antennae, characterized by their power parameter only. This parameter define the radioelectric field $F_m^{a_i} = f(Fade_m^{a_i} Power^{a_i})$ that each mesh receives from every surrounding antenna a_i. The goal of antenna parameter setting is to determine the values of the parameters in order to obtain the best service quality of the land.

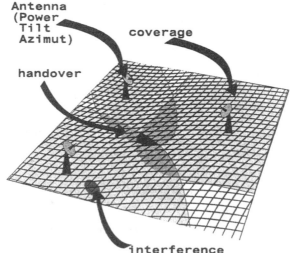

Fig. 1. the antenna parameter setting problem.

In fact, the term service quality includes several aspects : coverage, traffic, interference and handover. Indeed, we want every mesh m to be covered, i.e. to receive a field greater to a quality gate $G_{quality}$ from an antenna a_i, enabling the communication with a mobile phone in the mesh :

$$Cov_m^{a_i} = \left[F_m^{a_i} \geq G_{quality} \right] \tag{1}$$

We also consider that an antenna a_i is interfered by another one a_j if the field received from the latter is greater than a sensibility gate $G_{sensibility}$, namely :

$$I_m^{a_i,a_j} = \left[Cov_m^{a_i} \text{ and } G_{quality} \geq F_m^{a_j} \geq G_{sensibility} \right] \left(G_{quality} \gg G_{sensibility} \right) \qquad (2)$$

The notion of handover is used to enable a mobile phone to go from an area covered by one antenna to an area covered by another one. A handover area is covered by two antennae and has to be neither too small, hindering the effective transition between antennae, nor too large, which would be useless. So, a mesh m establishes a handover relationship between two antennae if covered by both and if the difference between the fields received is under the handover gate $G_{handover}$:

$$H_m^{a_i,a_j} = \left[Cov_m^{a_i} \text{ and } Cov_m^{a_j} \text{ and } \left| F_m^{a_i} - F_m^{a_j} \right| \leq G_{handover} \right] \qquad (3)$$

These notions are intuitively illustrated by figure 1.

2.3   Goals and constraints.

We stand from a strictly monocriterion point of view and only try to optimize coverage, considering handover and interference as constraints. We define them as follows :

coverage goal : maximize $\sum_i NC^{a_i}$ the number of meshes covered by an antenna

a_i.

handover constraint :

$$CH^{a_i} = \begin{cases} \text{if } TH_{min} \leq NH^{\cdots} \leq TH_{max} \text{ then } 1 \\ \text{else } 0 \end{cases} \quad (\forall a_j \in A, j \neq i) \qquad (4)$$

the proportion NH^{a_i,a_j} of meshes covered by an antenna a_i and in handover with any other a_j must be limited.

interference constraint :

$$CI^{a_i} = \begin{cases} \text{it } NI^{\cdots} \leq II_{max} \text{ then } 1 \\ \text{else } 0 \end{cases} \quad (\forall a_j \in A, j \neq i) \qquad (5)$$

the proportion NI^{a_i,a_j} of meshes covered by an antenna a_i and interfered by any other a_j has to be limited.

3   Evolutionary Algorithms approach.

Several methods have been applied to problems from the field of radiomobile networks, such as the APSP. EA, which include genetic algorithms (GA) and evolution strategies, have shown to offer good performances on such cases [15].

As other metaheuristic methods, GA are domain independent methods, characterized by their relative easiness of implementation and robustness for a large scale of applications [1], which require no or little specific domain knowledge. The main difference between MAS and metaheuristics-based methods [11] is that they rely on a fitness function, shifting the emphasis from « how to make a solution better » to « how to define a good solution ».

As a GA, by its very nature, does not require any knowledge about the solving mechanism [7], using one as a prototyping tool thus enables us to focus on the problem in order to learn about the exact nature of solutions. Heuristics can then be extracted in order to be used later on for the development of the MAS.

3.1 Genetic modeling.

The design of a GA depends on the following points : genetic coding, definition of an objective function, and definition of the crossover, mutation and selection genetic operators. The definition of the genetic coding comes from the previous model. A solution to the problem is defined by the state of all the antennae, i.e. their respective parameters of power, tilt and azimuth. As we rather stand from an evolutionary point of view, we do not make any difference between phenotype and genotype, and define the chromosome manipulated by the algorithm as the aggregation of the antennae parameters. We define a phenotype constituted from the three parameters of each antenna, as in the figure 2.

Antenna #1			Antenna #2				Antenna #n		
power	tilt	azim	power	tilt	azim	...	power	tilt	azim

Fig. 2. genotype and phenotype.

In the same manner, the objective function is defined from the criteria characterizing the problem. The goal is to maximize the coverage of every antenna and to satisfy the constraints of traffic, interference and handover, given the relationships between antennae. The quality of a solution (representing the configuration of the network as a whole) is logically defined as the sum of the quality of each antenna.

$$quality_{global} = \sum quality^{a_i} \ \left(\forall a_i \in A\right) \tag{6}$$

The objective score for an antenna has to avoid the definition of a multi-criteria score bearing on coverage, interference and handover as well. Therefore, it represents both the quality of the antenna coverage and the discrimination of the violation of constraints. It is given by the following equation :

$$quality^{a_i} = NC^{a_i} - K_i \sum \Delta_{int\,erf}^{a_i a_j} - K_h \sum \Delta_{hand}^{a_i a_j} - K_t \sum \Delta_{traf}^{a_i} \quad (\forall a_j \in A,\ j \neq i) \quad (7)$$

$$with\ \Delta_{int\,erf}^{a_i a_j} = NI^{a_i a_j} - TI_{max}\ if\left(NI^{a_i a_j} > TI_{max}\right)$$

$$with\ \Delta_{hand}^{a_i a_j} = \left|NH^{a_i a_j} - TH_{max}\right|\ if\left(NH^{a_i a_j} > TH_{max}\right)\ or\ \left|NH^{a_i a_j} - TH_{min}\right|\ if\left(NH^{a_i a_j} < TH_{min}\right)$$

$$with\ \Delta_{traf}^{a_i} = NT^{a_i} - E^{a_i}\ if\left(NT^{a_i} > E^{a_i}\right)$$

with K_i, K_h, K_t, great positive integers, to heavily penalize solutions violating constraints, in order to forbid them. This defines a strictly monocriterion objective function which gets rid of solutions violating constraints. The principle of this objective function can be represented as in figure 3 :

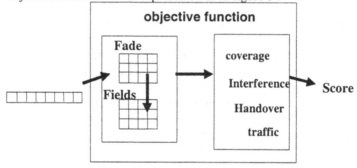

Fig. 3. objective function.

From the genotype, that is from the antennae parameters, and given the corresponding fading matrix, we compute for each antenna the resulting field matrix. We can then deduce the coverage, the interference and handover constraints, and compute the antenna score and thus the final score of an individual.

Fig. 4. typical solution found by our GA.

The GA evolves good solutions with a population size of 100 and number of generations of 200. Figure 4 shows an example subnetwork found by the GA once the population has converged and there is no improvement any more. Figure 5 presents the population average fitness and best individual fitness, which hopefully decrease and converge over a quite small number of generations. However, considering that we rather use this GA as a prototyping tool, we do not need to wait for a converged population. Indeed, a limited run of the GA, with low generation number (30) and population size (30) is sufficient to observe meaningful features in the evolution of solutions. This allows faster behavior study of the GA and faster prototyping.

The behavior of the GA, that is its solving mechanism, is conditioned by the various elements defining the evolutionary process and whose main component is the fitness function. The fitness function of a GA defines how fit individuals are according to the problem at hand and thus defines what type of solutions are to be found. In the case of the APSP, it is both characterized by constraint handling parameters, defined according to the optimization goals, and by network parameters such as constraint gates, which have an indirect influence on it.

The behavior study of the GA thus consists in back and forth observing resulting solutions and tuning these parameters. By examining the characteristics of solutions - even early ones, as stated before - it is possible to determine how the GA behaves, that is how solutions are evolved. This enables to extract heuristics about the solving mechanism, which can then be used for the MAS.

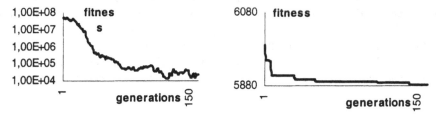

Fig. 5. population average fitness and best individual fitness.

3.2   Genetic behavior.

The heuristics were extracted when trying to solve problems related to the handover constraint : the inadequate shape of the constraint space and repartition of individuals in this space. Indeed, the population was dominated by low handover individuals, far more likely to appear than high handover ones. This situation produced a genetic drift leading to unacceptable solutions and was solved by modifying the fitness function. These heuristics can be summarized as follows :

- over-limit and under-limit handover constraints must be differentiated. The under-limit constraint then acts as a very hard constraint ensuring handover wherever the over-limit one plays a softer role, similar to interference and limit the amount of handover.

- solutions with high handover have to be favored. This allows to prevent the over-dominance of low handover individuals and implicitly respond to the coverage maximization goal.
- the handover constraint has to be privileged to the interference one. Indeed, for high handover individuals, handover satisfying solutions are to be preferred to interference satisfying ones, because closer to ideal solutions.
- the reaction to a constraint violation have to be proportionate with the violation, in order to get back to acceptable solutions.

4   Multi-Agents approach.

The reasons motivating the use of MAS for optimization problems from the field of radiomobile networks can be stated from two different perspectives. From a design point of view, a distributed approach reduces the complexity of modeling, as it deals with both simpler and natural entities of the problem [6]. From an optimization point of view, it gives the possibility of obtaining solutions presenting some features particularly interesting for cellular networks : evolutivity and flexibility [9].

However, MAS used for problem solving lack some natural combinatorial exploration power and built-in solving mechanism that have optimization techniques such as GA, successfully tried above on the APSP. This drawback should normally be counterbalanced by the use of heuristics to define the agents behavior, facilitated by the natural modeling of the problem and experimentation process.

In the case of the APSP, this could not be done from the start because of insufficient knowledge about the problem and led to the use of a GA as a prototyping tool. Some heuristics have been gained through this means and can then profit to the MAS.

4.1   Agent modeling.

The goal is to give a description of the problem, exhibiting the existence of agents, whose interactions lead the system as a whole to an emergent state [8], corresponding to the desired state. The system as a whole is composed of the meshes and the antennae. We consider the set of meshes as the environment of the MAS and the antennae as the agents perceiving this environment and acting on it. We precise the attributes of meshes and antennae, the events resulting in their reactions, and the actions they take.

4.1.1   Meshes

We consider the set of meshes as the environment of the system. Each mesh is characterized by the radioelectric fields from every antennae and by its corresponding status (coverage, handover, interference). Its goal is to define its own status and inform the antennae involved, enabling them to define coverage, handover and interference areas through their actions.

<u>attributes</u> : field $F_m^{a_i}$ and status of the mesh : (coverage, handover, interference).

<u>events</u> : antennae actions, that is modification of the fields.

<u>action</u> : actualize status and inform antennae.

The behavior of a mesh is thus defined in the following manner :

```
if  F^a_m > G_quality  then
            inform(a, coverage)
if  F^a_m > G_quality  and  F^b_m > G_quality  and  |F^a_m - F^b_m| ≤ G_handover  then
            inform(a-b, handover)
if  F^a_m > G_quality  and  G_quality ≥ F^b_m ≥ G_sensibility  then
            inform(b, interference(a))
```

4.1.2 Antennae.

We define the antennae as the agents of the system. They react to the stimuli from the meshes, and, according to the status of the latter, check if they respect constraints. When violating a constraint, they act consequently in order to get back to an acceptable state. In the contrary, when no constraints are violated, they exhibit some proactivity. They act according to their default behavior responding to their goal and try to maximize their coverage.

<u>attributes</u> : parameters $Power^a$, $Tilt^a$, $Azim^a$, and variables used for constraints checking NC^a, NI^a, NH^a.

<u>default behavior</u> : increase coverage.

<u>events</u> : reception of status change message from meshes.

<u>actions</u> : checking constraints and setting parameters to modify the environment.

The agent behavior can be defined as follows :

```
if   ∑(m∈MC^a) E_m > E^a  then
            adjust parameters (increase/decrease)
if   [ NI^{a,i}/NC^a > TI_max ]  then
            adjust parameters (inc./dec.)
if   NH^{a,i}/NC^a < TH_min  or  NH^{a,i}/NC^a > TH_max  then
            adjust (inc./dec.)
            else increase coverage (default behavior).
```

The definition of this MAS components is based upon a reactive cooperation, implicitly occurring between antennae through their actions in the environment - that is on the meshes - as if by the means of wavefields. The definition of this component allows the solving mechanism to the APSP through their coordination, based upon a reactive cooperation. We now precise the agent behavior ensuring the apparition of such reactive cooperation mechanism.

4.2   Agent behavior.

The agent behavior is built from the intuitive behavior scheme coming from the APSP model. In order to obtain a coherent and effective system behavior, they however have to take into account the heuristics extracted with the GA : separating the handover constraint, favoring high handover configurations, balancing handover and interference, and eventually reacting proportionately to constraint violation.

We separated reactions about the handover constraint to establish two distinct behaviors : the first corresponding to too much handover and the second to a lack of it. Moreover, we privileged handover by initializing the simulation in a state where antenna power is high. They establish handover but are hindered by interference. They then try to lower interference and eventually reach an intermediate state (where both interference and handover are respected) from above, that is with high power. The maximization of coverage is thus implicitly favored. However, we observed that the stable solutions we reached could be better in terms of coverage. This could be solved by taking into account the last heuristic and making the reaction proportionate to constraint violation.

The remaining issue is the adjustment of the parameters when violating constraints. We have chosen a method inspired from the optimization technique of simulated annealing [11], which travels through the search space with a given probability of degradation, and progressively decreases this probability. In our case, the parameter adjustment is chosen randomly, according to a decreasing probability. However, conflicts due to constraints violations occurring between antennae could result in their simultaneous reaction. This would tend to end in the opposite situation and actually result in a cyclic behavior of the system. This is avoided by the satisfaction and agitation reflexes of the agents. Indeed, an antenna will see its agitation, or probability to react, decrease with time, and once satisfied, i.e. when all its constraints are satisfied, will cease to act. So, through time, an antenna will react less often to stimuli and when repeating the same behavior will react more softly and eventually cease to react when satisfied.

This issue of behavior design would be different, should we consider directional antennae. Indeed, the management of the three parameters (power, tilt and azimuth) for such antennae would imply the definition of complex behaviors, and take into account the relative positions of antennae and meshes.

We have used the Actalk kernel [2], which stands for « Actors in Smalltalk », as a basis to develop our multi-agent system [3]. It defines a set of Smalltalk classes enabling the modeling of active and autonomous objects communicating via message passing, and thus appears sufficient to define (reactive) agents, as those we consider. Even though Actalk enables the definition of asynchronous agents, the simulation of the multi-agent system is implemented in a synchronous way. Indeed, once every mesh has informed the responsible antennae of its own status, all antennae react simultaneously according to those stimuli.

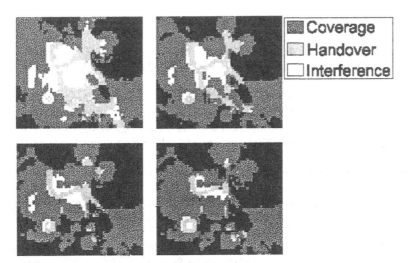

Fig. 6. example evolution sequence of the system.

Figure 6 shows an example evolution sequence of the multi-agents system while it is running. The MAS eventually ends with an acceptable solution, which satisfies every constraint. However, this could be better, should the last heuristic be incorporated in order to obtain acceptable solutions faster, and should the simple reactive behavior be refined and tuned during an experimentation process. This shows that MAS techniques can easily be applied to the field of radiomobile networks, and demonstrate acceptable results. Work should also be done in the future regarding the suitability of MAS solutions to constraint and network modifications.

5   Conclusion.

This paper has presented an approach based upon the use of an evolutionary algorithm as a prototyping tool, to develop and refine a MAS for problem solving. It demonstrated the feasibility and interest of such an approach for complex problems such as antenna parameter setting. Moreover, it has advocated the use of MAS techniques for the field of radiomobile networks.

Future work on the joint use of EA and MAS remains to be conducted regarding the possibility of reciprocal contribution between the two techniques. In this perspective, knowledge gained at studying the agent behavior could also be used to improve the GA through its genetic operators and especially mutation. This would open the way to a global joint MAS/EA approach to problem solving.

6   Bibliography.

[1] Beasley D, Bull D.R., Martin R.R. (1993). *An Overview of Genetic Algorithms.* University Computing.

[2] Briot J.P. (1988). From Objects to Actors : Study of a Limited Symbiosis in Smalltalk-80. *Research Report RXF-LITP*, n°88-58.

[3] Cardozo E., Sichman J.S., Demazeau Y. (1993). Using the Active Object Model to Implement Multi-Agent Systems. In *Proc. of the 5th IEEE Int. Conf. on Tools with Artificial Intelligence.*

[4] Durfee E.H., Rosenschein J.S. (1994). Distributed Problem Solving and Multi-Agent Systems : Comparisons and Examples.

[5] Ferber J. (1995). *Les systèmes multi-agents : vers une intelligence collective.* InterEditions.

[6] Ferber J., Jacopin E. (1991). The framework of EcoProblem Solving. *Decentralized AI, vol. 2.* Elsevier.

[7] Goldberg D.E. (1994). *Genetic Algorithms in Search, Optimization and Machine Learning.* Addison-Wesley.

[8] Groupe de travail "Collectif" IAD/SMA de AFCET/AFIA (1997). Emergence et Systèmes Multi-agents. In *Actes des 5ème Journées Francophones d'Intelligence Artificielle et Systèmes Multi-Agents*, pp 323-341. Hermes.

[9] Guedira K. (1993). MASC : une approche multi-agents des problèmes de satisfaction de contraintes. Thèse de Doctorat, Ecole Nationale Supérieure de l'Aeronautique et de l'Espace.

[10] Guisnet B. (1996). La propagation pour les services de mobilité. *Les Communications avec les Mobiles.*

[11] Hao J.K., Galinier P., Habib M., Méthodes heuristiques pour l'optimisation combinatoire et l'affectation sous contraintes. In Actes des 6èmes journées nationales, PRC-GDR IA, pp 107-144. Hermes.

[12] Labidi S., Lejouad W. (1993). De l'Intelligence Artificielle Distribuée aux Systèmes Multi-Agents. *Rapport de recherche n°2004.* INRIA.

[13] Magnin L. (1996). Modélisation et simulation de l'environnement dans les systèmes multi-agents - Application aux robots footballeurs. Thèse de doctorat. Université Paris VI.

[14] Matthews D. GALIB a C++ Library of Genetic Algorithm Components, http://lancet.mit.edu/ga/.

[15] Renaud D., A. Caminada (1997). Evolutionary Methods and Operators for Frequency Assignment Problem. *SpeedUp Journal*, vol. 11, nr. 2, pp. 27-32.

Scalable Service Deployment on Highly Populated Networks

Luis Bernardo and Paulo Pinto

IST - Instituto Superior Técnico, Lisboa Portugal
Inesc, R. Alves Redol, 9 P-1000 Lisboa Portugal
Phone: +351.1.3100345   Fax: +351.1.3145843
{lflb,paulo.pinto}@inesc.pt

Abstract. Very large networks with thousands of applications and millions of users pose serious problems to the current traditional technology. If synchronised client behaviour exists, then the limitations are even stronger. Cooperative agent systems can be a suitable technology to answer to these requirements if some aspects of the architecture are carefully designed. Particular attention should be given to the trader component and to the dynamic behaviour of the applications in response to client demand. This paper proposes a cooperative mobile agent system with a very dynamic and scalable trading service. The system is designed to allow applications to deploy servers onto the network to respond to demand making them self-configurable. Sets of simulations were performed to study the dynamic behaviour of the overall system, and identify the relevant tuning parameters.

1   Introduction

One of the problems facing designers of client/server applications, which are deployed on large-scale networks with millions of users, is the dimensioning of the server entities. Moreover, a characteristic common to some of the applications is the possibility of a synchronous pattern on the behaviour of the clients, producing peaks of traffic on servers. Examples of such applications are easy to define and can be tele-voting, tele-shopping, real-time sports brokering, stock brokering or applications based on interactive TV interfaces. These applications will involve a large number of clients, that may produce a burst of requests after a relevant event (e.g. a team scores a point, the announcement of promotional prices, or a deadline is approaching). It is essential that the application must satisfy the service response time requirements even under these extreme conditions. Therefore, both its static and dynamic behaviours need a careful design.

An efficient design choice is the possibility of launching a variable number of servers to process client requests in parallel (assuming that the nature of the service allows server mobility). A static approach to the problem, using a fixed number of servers and conventional traders, leads to inefficient resource usage solutions: either the number of servers is insufficient or there is over-dimensioning of the servers deployed. Most of the traditional dynamic solutions use a strictly system-based approach, instead of a per-service solution: they rely on system components to balance the requests amongst a constant pool of machines [2], [4], [6], [9], possibly deploying new servers. However, a worldwide service must not be based on a limited set of machines. Servers must be spread worldwide near the client's location to avoid bandwidth bottlenecks. The dynamics on the client's location over time will guide the server's deployment, resulting in lower client-server communication delays [20]. Such requirements can be achieved with any system providing remote object creation with state initialisation (such as factory objects or ORB implementation repositories [16]). However, the distribution of the server implementation would have to be made on all possible platforms. Mobile agent platforms provide a simpler way to deploy personalized services since all servers are obtained by cloning an initial replica.

The proposed system is based on a mobile agent system platform. Its architecture does not conflict with the standardization efforts of OMG [17], or some of the available mobile agent systems ([10], [15], etc). The main differences are the requirements of the trader (called here location service).

As it is described in this paper, our system assumes a scenario where clients look for a precise service. The system does not support service discoveries based on characteristics [13] (price, availability, etc.). Clients look for applications using a unique application name that is resolved to a server reference by the location service. Each application server sets an area of the network for its service to be known (server domain). Then, the location service splits the network dynamically into areas depending on the server population, the server domains, and the location server's own load. The location service has a new scalable algorithm to adapt to its own load (lookup requests) and to advertise the services of the servers it is responsible for. An overview of the location service is given in section 2.

Server mobile agents are autonomous on their control over server deployment. Servers use the dynamic topological information of the location service, the client load, and the overall situation of servers that belong to the application they are serving, to control the deployment of new servers and adapt precisely to the client load. The aim is to produce a highly flexible and scalable system that can support millions of interactions maintaining the desired quality of service. The proposed adaptation algorithm works as follows: each server monitors its client load and compiles the domains of the clients. When a peak on the load occurs, the server agent reacts creating clones, and deploying them based on the client origin information. A market oriented algorithm is used to reverse the process - lower the number of servers, when the client load decreases.

The study of the dynamic nature of the system covers too many aspects to be described in a single paper. In this paper we present a brief overview of the system components, the server deployment algorithm and a study of the dynamic behaviour of servers in face of a rising client demand (the adaptation to the worst case). The location service, which is itself a special dynamic service with its own load adaptation algorithm, will be covered on another paper.

2   System Overview

The network provides a ubiquitous platform of agent systems, in which any agent (server or client) can run. Each agent system is tied to a location server (running locally or on another nearby system), where all the interfaces of the local agents are registered. This location server is connected to others to offer a global location service.

When a client searches for an application name, the location service helps in the binding process (the association to a server) directing it to the nearest server. If the location server knows more than one server, it will do splitting of the client traffic. If it knows that a new, and closer, server was created it will start using the new one, and propagates this information. When a client comes for resolution, it will get the best answer for that moment. The balance between the number of servers, clients and location servers acts as a general load balancing mechanism in the system.

An application may scale and respond to peak conditions by deploying new servers, and thus increasing its total server processing capacity. The effectiveness is conditioned by the amount of time a server is inactive during the duplication process (which must be compared with the response time required by clients); and by the extra overhead to maintain consistency of the shared data due to the existence of a new server. The intra-server synchronisation is specific to each application (a simple placement of an order would not need such logic).

2.1   Location Service

The location service is one of the major players for scalability. Its requirements include: the necessity for fast updating during the creation of a server clone, the propagation of frequent updates due to server migration, and the dynamic nature of the information in the overall system (based on dynamic server domains). These requirements introduce a high overhead, which invalidates some of the current technical solutions, based on static hierarchical systems. Particularly:

- The use of cached values at remote nodes makes a fast change on the configuration information impossible [1].

- The use of full path names which define completely the search path [1], [11], [18] requires the use of a "home" server to keep all service related information (creating a bottleneck), or the backpropagation of a modification through the entire location network (making a update a costly operation).

None of these techniques must be present on a scalable and highly mobile system. First of all, the application names must be flat ([21] reached a similar conclusion). Secondly, the search path, which is now independent from the name structure, must be performed on a step-by-step basis, through a path of location servers where each one contains routing information indexed by the application name. Thirdly, this step-by-step path should be tuned by the load and characteristics of the overall system.

The routing information for the path is based on service references. References are either the full application name or incomplete information about the application just to direct the search to another location server. The objective is to keep references small and easy to update.

One important feature is how the location service scales to a large population. We use a mixture of meshed and hierarchical structure, where location servers at each hierarchical level interact with some of the others at that level and (possibly) with one above. Higher hierarchical levels always have incomplete information about the available services. Additionally, the hierarchical structure and the scope of the mesh change dynamically according to the load of the system, and to the size of the server domains.

The size of the server domain is service specific. For instance, a car parking service would simply advertise on the surroundings of each car park, while a popular lotto broker service would advertise on a broader range (pricing schemes could be a deterrent to artificially large domains). The structure of the location service hierarchy

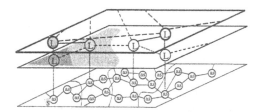

Fig. 1. Location Service Network Model

(hierarchical levels and meshes) will vary to gather the necessary number of agent systems the server wants in its domain. Higher hierarchical levels offer a broader, but less detailed vision of the services available. On the other side, clients control their search range. Due to lack of depth, or incomplete information, resolutions can fail, and a deeper search must be tried.

Figure 1 shows an example of the location service. The lowest plane, the agent system plane, has sets of agent systems forming meshes. The second and third planes show two hierarchical levels of the location service (with meshes in each one, and upward connections not designed). In this particular case, a reference to the server 's' is known completely on all agent systems at the server location's domain (darker grey area). The lighter grey area represents the scope where incomplete information is known (the reference to the location server associated with the agent system where the server is running).

3 Server Deployment

A good measure of the quality of service for this system is the global server response time to client requests. For each application, this time must be controlled within specific bounds. It includes: the time to resolve the application name to a server reference, plus a waiting time on the server due to client load, plus a service time dependent on the application (which depends on whether it is a single RPC or a session, the overheads for distributed data consistency, etc.).

The controllable system parameters are the number of servers deployed, their location and each of the server domains. The main parameter will be the number of servers, which defines the total available server processing capacity. At some instant, the "processing capacity ratio" (ratio between the number of servers deployed and the number of clients entering the system per time unit multiplied by the average service time) quantifies the availability of processing resources to satisfy the demand of new clients. The variation of the waiting time depends on the value of the processing capacity ratio (PCR) and on the distribution of clients per server. It gets higher when the PCR is below one, and gets lower otherwise (assuming a completely balanced system). As clients are bound to servers based on the distance and on the relative importance of the server (broader server domains get more clients), some unbalancing can exist depending on the relative distribution of clients.

3.1 Deployment Algorithm

The number of servers and their location could be configured if the service usage peaks and the origin of the requests could be anticipated. Unfortunately, for most of the applications, there is only a vague forewarn of how much the "peak load" may be, or when it will happen.

The proposed algorithm may be used with various client-server interaction methods [3]. However, it provides the best results when the interaction methods allow servers to know the precise number of clients using a specific server (server's load). This knowledge can be obtained from the client pending requests on the server's input queue (either session connections or RPC invocations). Our algorithm uses this information as well as local machine load, ignoring the number of clients that are trying to reach the server and fail. It also compiles the clients' origins and keeps a statistic indexed by the location server identifiers.

When the client load goes above a top threshold value, the server creates and deploys a new server. This action is isolated. It does not involve interaction between application servers. The new server's location server is selected amongst the most frequent sources of agents (local or not). The new server is created on an agent system picked from a list returned by the selected location server. The new server will only be completely available to run client requests after T_{clone}, which is the time to create a clone on the remote agent system, plus the delays at the location service (dissemination of the new clone's server reference). During this period, new clients continue to bind to an already overloaded server. So, the triggering mechanism of the top threshold value is disabled for a duration dependent on T_{clone}. When the total number of pending clients is known, a temporary increase on the top threshold value can be made to include an estimate of the number of clients that would be processed by the new server during the disabled period.

If the demand is very high, and the waiting time exceeds largely the response time, the server can unbind some of the clients, or can mutate itself (i.e., close the old interface and create a fresh one). The number of pending clients (if available) or the number of consecutive load overflows are used to detect such conditions. Unbinds and mutation will force a new resolution phase for all waiting clients and a redistribution of the clients for the available servers.

A market-based control technique is used to reverse the algorithm. When the load goes below a bottom threshold, the server sends a message with a "request for bids" to all the neighbour servers (within a distance range in the location network), requesting one of them to take its place. Requested servers will answer with a bid, stating if they can expand their domain, and stating their load. The requester will wait for answers during a time interval, and selects the best bid within the minimum distance with the minimum load, afterwards. Notice that the value of the time interval is not as relevant for the system performance as in other market oriented systems (e.g. [6], [14]), where the system response time is directly related to this value. The bottom threshold is only enabled for dynamically created servers. Permanent servers (created by service provider's specification) live as long as service providers want.

The presented algorithm scales to broader networks than alternative approaches (transaction managers [4], load balancing systems [9], broker/matchmaker agents [8], or market oriented systems [6], [14]) since the adaptation to overload conditions is based on an isolated algorithm. However, it may cause the deployment of a higher number of servers since the location service balance client requests amongst servers known locally not taking into account the server's instantaneous load. The equilibrium between the client load and the number of servers deployed is achieved at limited ranges, instead of at a global level.

4 Dynamic Behaviour

The analysis of the dynamic behaviour of the system was conducted using a simulation model. The set of tests presented focused on the adaptation to a constant demand from clients. The simulation model runs all the services described so far but the interference of some algorithms was avoided. The effect of the location service was reduced by setting a very low-resolution time (compared to the application service time), and by disabling the dynamic change of the hierarchy. Nevertheless, it still runs the application name distribution algorithm, which introduces a delay between the deployment of a new server and the stabilization of the location service information (proportional to the transmission delay between servers). The effects of the variation of the latency on the communication between agents were disabled, by setting it to a constant value. This very symmetrical scenario produces highly synchronized reactions on servers, but it is clearly the worst case.

With respect to the adaptation algorithm, we studied three alternatives for redistributing clients when a client load peak happens. The first is to mutate the server completely, to even avoid those clients that made the name resolution but are not yet in the queue (*total unbind*). The mutation is executed after the new clone becomes operational if the queue is still overloaded. When several clones are created, deactivation is delayed until the last server starts. The second is to unbind those clients in the queue whose waiting time exceeds *Timeout* (*partial unbind*). The third is to serve all clients not doing any unbinding (*none*).

4.1 Simulation Environment

The simulator was developed using the "Discrete Event" model on the Ptolemy [19] simulation system. All tests were conducted on a network presented in figure 2, with 132 agent systems and 19 static location servers. Results were collected at the end of each measuring interval of 0.5 units of simulation time. The duration of each simulation was 30 time units (*tics*).

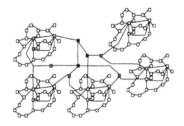

The simulation assumes an atomic interaction between the client and the server agents. A client is born and lives until it can make an invocation to the server. The total number of clients accessing a server is

Fig. 2. Simulated meshed network

supposed to be known. Our main results are the client's lifetimes, which are the overall application response times.

The application and location service servers are modelled by a queue defined by a service time probability function, T_s and T_L respectively. For all the experiments reported in this paper, T_L and T_s were deterministic functions with the values 0.001 and 0.1 tics respectively. The transmission time was set to 0.0001 tics. Servers use the number of requests in the queue as an indication of client load. The top threshold value is called "Maximum Client Queue Trigger Level" (*MaxCliQ*) and the disabled time after a clone is launched is 1.5 times T_{clone}.

The client creation is defined by an inter-client generation statistic, and clients are deployed with a uniform distribution to a set of 125 agent systems. The inter-client deployment statistic is defined by a uniform distribution on the interval [0, 2/ClientLoad], where *ClientLoad* defines the average number of clients that enter into the system during a time unit.

4.1.1 Results

The simulator measures the client's lifetime, the number of clients and servers, and their state. Therefore it is possible to have an evolution over time of the averages on the measuring intervals. Figure 3 shows the evolution of: *New Clients*, the number of clients which entered the system; *Unbinds*, the number of clients unbound during the interval; *Pending Clients*, the number of clients waiting on queues (of both application and location servers); *Ending Clients*, the number of clients which terminated during the interval; and *Processing Capacity*, the number of servers times the service time (which measures the number of clients which can be processed on the interval). The second graphic shows the evolution of the average global response time per client measured on each interval, represented by *TT* (Total Time). The curves were measured with ClientLoad= 250 clients per tic (125 new clients per measuring interval), using partial client unbind with $T_{clone}= 1$, MaxCliQ= 15 clients and Timeout= 1.5. It is also possible to calculate: the average value, TT_{avg}; the worst client life value, TT_{worst}; and the time value that includes 95 percent of all clients' lifetimes, *TT95*.

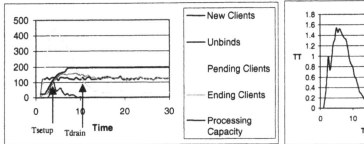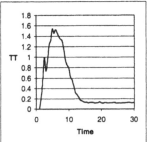

Fig. 3. Service Response – evolution on time of the number of clients and of the average total delay

As soon as client requests start (at tic 1), the number of pending clients grows and the system starts to adapt. At T_{setup} the processing power deployed is already enough for the client load. After T_{setup} the number of pending clients starts to decrease. TT continues to grow just for a short while after this point (the curves are almost equal for the experiment of figure 3 because unbind was used). T_{drain} measures the instant when the number of pending clients is below arrival rate for the measurement interval (new clients). It is clear how the system gets stable with a very low and constant response time. All the presented values are averages of at least two experiments, and can be scaled to an arbitrary system using the relation between the service time values.

4.2 Results with Weak Inter-server Synchronisation

Figure 4 present the global response time histograms for the three unbind methods. The resulting TT_{avg} and standard deviation values for partial, total and no binds are respectively: 0.52 ± 0.81, 0.42 ± 0.99 and 0.93 ± 1.49 tics. Both unbind methods improved the system response time, with a better overall performance for the server mutation. However, the measured peak number of unbound clients is very high (274 during a measurement interval) compared to unbound clients with partial unbind (98) and to the average number of new clients (125), resulting in PCR values of respectively 1.64, 3.86 and 1.38 (41, 97 and 35 servers). A minimum of 25 servers was needed to satisfy the new client requests.

A particularly nasty side effect of the total unbind procedure for our very symmetrical scenario is the following: new servers receive a "peak" of both redistributed clients and new ones, which may be higher than MaxCliQ, generating unnecessary new clones. The partial unbind method is used for the remaining experiments, because it is the one which offers the best trade-off between the deployment cost and the measured response times.

38

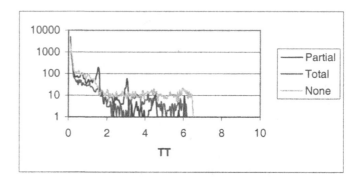

Fig. 4. Histogram[1] of response times for: partial client unbinding (with Timeout=1.5), server deactivation (total), and no unbinding, ClientLoad=250, MaxCliQ=15, five starting servers

4.2.1 Parameters Configuration

For each application, the values for service time and T_{clone} will influence the minimum achievable response time. Figure 5 shows the system performance for five values of T_{clone}, when a peak of 250 clients per tic is injected on a system with five initial servers (initial capacity of 50 clients per tic). As expected, T_{setup} depends strongly on T_{clone}, resulting in more pending clients for higher T_{clone} values. Consequently, client lifetime (TT95 and TT_{avg}) increases, and a higher number of servers is deployed (PCR). For the smallest value of T_{clone} (0.2) it is still noticeable a drain time, which exists even for T_{clone} = 0, due to buffering on the server queues (MaxCliQ is above zero).

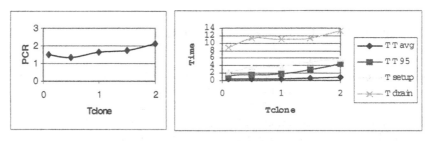

Fig. 5. T_{clone} sensibility with ClientLoad=250, MaxCliQ=15, Timeout=1.5, five initial servers

Several parameters can be set to optimise the algorithm performance. Figures 6a, 6b and 6c show the distribution of TT95, TT_{avg} and PCR for a set of 25 pairs of values for MaxCliQ and Timeout, with ClientLoad= 250 clients per time unit, T_{clone}=1 and five initial servers.

---

[1] 0 values were converted to 1 to allow a logarithmic representation of the number of samples

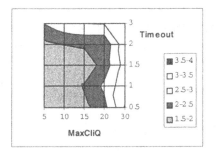

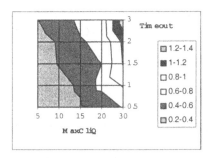

Fig. 6a. TT95

Fig. 6b. TT$_{avg}$

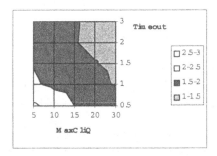

Fig. 6c. PCR

The Timeout value controls the redistribution rate between servers, and MaxCliQ controls the trigger value to launch new servers. Together, they control the system response speed to a peak of requests. The figures show a strong dependency on both parameters with an interesting trade-off: if the response time has to be very low, then PCR will rise (generating too many clones). The configuration must be defined considering the application requirements. For instance, a fast response (TT$_{avg}$ < 4 * T$_s$ = 0.4 tics) implies PCR > 2.5. The fastest measured response time (MaxCliQ=5 clients, Timeout=1 tic) had 99% of clients with a lifetime less than 2.725 tics. The best possible response time is T$_{clone}$ plus service time (1.1), with the creation of one server for each waiting client. For Timeout values below T$_{clone}$, clients get unbound before a new server is deployed, which originates a high number of creations, and the measured high value for PCR.

Figures 7a, 7b and 7c show the distribution of TT95, TT$_{avg}$ and PCR for twenty configurations of different ClientLoad and initial number of servers, with MaxCliQ=15, Timeout=1.5 and T$_{clone}$=1.

The results show a minor increase of the response times (TT95 and TT$_{avg}$) and of PCR (the ratio between number of servers deployed and the minimum necessary) compared to the increase on ClientLoad (800%), which proves the algorithm scalability. If the total number of clients bound to a server is not known then the response would be slower, and the ratio would decrease. However, it would always be low, since the number of servers deployed grows exponentially. The initial number of servers has a great influence on the three parameters, specially for lower ClientLoad values. When the load is high, the adaptation is done quicker due to a flooding of servers and the PCR gets higher (over-deployment of servers). In this case, the initial number of servers becomes irrelevant. It is interesting to note that for ClientLoad = 125 and five initial servers the system moves smoothly and the response times are very acceptable. This proves that some load expectations from the application can produce very stable adaptations.

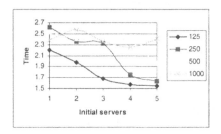

Fig. 7a. TT95

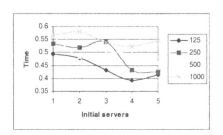

Fig. 7b. TT_{avg}

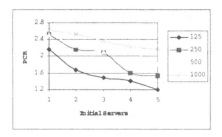

Fig. 7c. PCR

Additional experiments using random distributions (exponential, normal and uniform distributions) with the same average for T_s showed that the results were comparable to the ones measured with the deterministic value. There were slight increases on the response times (less than 20%) and on the number of servers deployed (less than 10%), inducing that the algorithm is sufficiently generic.

4.3   Server Deployment with Strong Synchronised Systems

Most services require some state synchronisation between servers. This will introduce a limit to the maximum number of clients (load) which can be processed per time unit. It is still possible to use the algorithm on these cases with minor corrections as long as the client load is below the maximum value supported. The average service time increases when a new server is created (and decreases when one dies). When the service time increases, fewer clients are serviced per time unit. In consequence, it takes less time to reach MaxCliQ clients waiting in the queue. The ratio between Timeout and average service time degrades until it possibly goes below 1, resulting on an explosion of server creation (a very high PCR). The algorithm was modified to avoid this effect: MaxCliQ and Timeout are incremented when the average service time increases and decremented otherwise. It lets the system adapt more slowly to peak loads.

We tested the approach on a system with a linear degradation for each server (which models a periodic synchronisation between the servers), with the service time given by the following formula:

$$\text{ServiceTime} = 0.1 \times \left(1 + \alpha \times \left(\text{NumberServers} - 1\right)\right) \tag{1}$$

Table 1 shows the measurements with ClientLoad=250, T_{clone}=1, MaxCliQ=15, Timeout=1.5 (initial values), and five initial servers. The columns "Servers for T_{setup}" and "Max PCR" present the number of servers needed to sustain the incoming clients and the theoretical maximum value[2] achievable for the rate between the maximum number of clients and ClientLoad (with infinite servers). For $\alpha \geq 0.04$ the system can not support the load of 250 clients per tic (MaxPCR is below 1).

α	Servers for T_{setup}	Max PCR	T_{setup}	T_{drain}	TT_{avg}	TT95	TT_{worst}	Serv. Dep.	PCR
0	25	∞	5	11	0.50	1.9	7.86	41	1.64
0.005	29	8	5	13.25	0.63	2.1	6.85	43.5	1.5
0.01	33	4	5	22.5	0.97	2.7	9.68	49	1.49
0.02	49	2	7.5		2.40	5.9	17.78	76	1.55
0.03	97	1.33	24.5		4.12	10.4	24.21	112	1.15

Table 1. Performance with synchronised clients

Increases to the value of α lead to reduced processing capacity gains for each new server added to the system, and in result, to slower responses to client request peaks, shown by the increase of TT_{avg} and TT95. For the two higher values of α, this effect even prevents the system from draining the clients accumulated during the deployment of servers (T_{drain} is above the simulation duration of 30 tics). The difference between α=0.02 and α=0.03 is that the number of pending clients is much higher for α=0.03 and the exceeding available processing capacity (PCR) is very low (15%), resulting in higher response times (well above the simulation duration). Without the algorithm modification the system deploys ("Serv. Dep.") 284 servers with TT_{avg}=1.1 and TT95=3.2 for α=0.02, and deploys over 500 servers for α=0.03.

5 Related Work

The design of scalable systems to support worldwide applications is addressed on [7], [21], where architectures for global location services and server replica co-operation are proposed. There are some differences regarding the location service but this is not

---

[2] $MaxPCR = 1/(0.1 \times ClientLoad \times \alpha)$

the main subject of this paper. None of them, however, handles the "client peak" invocations due to some external and uncontrolled event.

Dynamic server replication systems are proposed on [2], [5] for optimising the bandwidth usage on the access to WWW servers. The client dissemination amongst server replicas uses a "master" server, with the resulting scale limitations.

[12] proposes a low level approach based on fine-grained objects. The scalability is compromised by the lack of an "anycast group address" and by the use of front end scheduling objects.

Market oriented based systems [6], [14] are proposed to bind client-server interaction. However, the communication of "bid" messages between each client and all the servers presents strong scale limitations.

6 Conclusions and Future Work

This paper presents a cooperative agent system that allows applications to scale to large networks with millions of users. The dynamic behaviour of the algorithm in face of a strong rise on client demand was studied and several conclusions were drawn with the experiments.

An overall conclusion is the suitability of such systems and algorithms to respond to "client peak invocations". Traditional systems do not scale and will create severe bottlenecks if used under these conditions.

The simulation results showed that it is possible to guarantee a limited application response time for 95 per cent of the clients if some applications parameters are known. Namely, the maximum values for the service time, for the clone deployment time (related to the transmission time of the server agent, the bandwidth and the computational resources of agent systems), and for the name resolution time. By a correct control on the number of replicas initially deployed and the correct setting of Timeout and MaxCliQ, an application may be ready to respond to a roughly predicted rise on the client demand. Even when the demand is not predicted the system reacts well, but sometimes, it can create too many servers due to a rapid increase on the offered capacity.

The algorithm is used in conjunction with a location service, which supports the scalable trading between clients and servers. The mutual relation between both mechanisms is a challenging issue and a good direction for further study.

This paper covered atomic interactions between clients and servers. Multi-invocation interactions can introduce other requirements to the algorithms and will be studied as well.

Client implementation is also a concern. Clients may communicate from their agent systems, or can also be implemented as mobile agents, migrating to agent systems nearer the server they want to use. It will be interesting to see the trade-offs between both approaches.

Acknowledgements

This research has been partially supported by the PRAXIS XXI program, under contract 2/2.1/TIT/1633/95.

7   References

[1]   Albitz, P., Liu, C.: DNS & BIND. O'Reilly &Associates Inc. (1996)
[2]   Baentsch, M., Baum, L., Molter, G., Rothkugel, S., Sturm, P.: Enhancing the web's Infrastructure: From Caching to Replication. IEEE Internet Computing, Vol. 1 No. 2, March-April (1997) 18-27
[3]   Baumann, J., Hohl, F., Radouniklis, N., Rothermel, K., Straßer, M.: Communication Concepts for Mobile Agent Systems. In: Mobile Agents - Proceedings of the First International Workshop on Mobile Agents (MA'97), Germany, April (1997) 123-135
[4]   BEA: TUXEDO White Paper. (1996) http://www.beasys.com/Product/tuxwp1.htm
[5]   Bestavros, A.: WWW Traffic Reduction and Load Balancing through Server-Based Caching. IEEE Concurrency, Vol 5 N 1, January-March (1997) 56-66
[6]   Chavez, A., Moukas, A., Maes, P.: Challenger: A Multi-agent System for Distributed Resource Allocation. In: Proceedings of the International Conference on Autonomous Agents, Marina Del Ray, California (1997)
[7]   Condict, M., Milojicic, D., Reynolds, F., Bolinger, D.: Towards a World-Wide Civilization of Objects. In: Proceedings of the 7th ACM SIGOPS European Workshop, Ireland, September (1996)
[8]   Decker, K.: Matchmaking and Brokering. In: Proceedings of the Second International Conference on Multi-Agent Systems (ICMAS-96), May (1996)
[9]   Deng, X., Liu, H.-N., Long, J., Xiao, B.: Competitive Analysis of Network Load Balancing. Journal of Parallel and Distributed Computing Vol. 40 N. 2, February (1997) 162-172
[10]   IBM Aglets Workbench - Home Page. http://www.trl.ibm.co.jp/aglets/
[11]   ISO/IEC: Information technology - Open Distributed Processing - The Directory - Overview of concepts, models, and services. ISO/IEC DIS 9594-1, ITU-T Rec. X.500. November (1993)
[12]   Kim, W., Agha, G.: Efficient Support of Location Transparency in Concurrent Object-Oriented Programming Languages. In: Proceedings of the Supercomputing'95, San Diego, December (1995)

[13] Lynch, C.: Network Information Resource Discovery: An Overview of Current Issues. IEEE Journal on Selected Areas in Communications Vol. 13 N. 8, October (1995) 1505-1522

[14] Mullen, T., Wellman, M.P.: A simple Computational Market for Network Information Services. In: Proceedings of the 1st International Conference on Multi-Agent Systems (ICMAS'95), San Francisco, June (1995) 283-289

[15] ObjectSpace Voyager V1.0.1 Overview. http://wwwobjectspace.com/voyager/

[16] OMG Inc.: The Common Object Request Broker: Architecture and Specification, Rev 2.0. July (1995)

[17] OMG Inc.: Mobile Agent Facility Specification. OMG Draft, October (1997) ftp://ftp.omg.org/pub/docs/orbos/97-10-05.pdf

[18] OMG Inc.: Trading Service. OMG TC Document 95.10.6, October (1995)

[19] Ptolemy project home page. http://ptolemy.eecs.berkeley.edu/

[20] Ranganathan, M., Acharya, A., Sharma, S., Saltz, J.: Network-aware Mobile Programs. Technical Report CS-TR-3659 and UMIACS TR 96-46, Department of Computer Science and UMIACS, University of Maryland, June (1996)

[21] van Steen, M., Hauck, F., Tanenbaum, A.: A Model for Worldwide Tracking of Distributed Objects. In: Proc. TINA '96 Conference, Heidelberg, Germany, September (1996) 203-212

Heterogeneous Multi-Agent Architecture for ATM Virtual Path Network Resource Configuration

A. L. G. Hayzelden and J. Bigham

Intelligent Systems Applications Group
Department of Electronic Engineering,
Queen Mary and Westfield College,
University of London.
London, E1 4NS.
{a.l.g.hayzelden, j.bigham}@qmw.ac.uk

Abstract. This paper illustrates a new approach to the problem of making a logical network resource configuration, adapt to customer utilisation. This is to be achieved through the use of distributed multi-agent based control techniques. The agents have goals derived from different quality metrics, which the network has to provide. In ATM networks a well designed Virtual Path Connection (VPC) overlay network tries to maximise the probability of being able to accommodate connection demand within the control plane for that particular time frame. We assume that there already exists a reasonable VPC route topology in operation. The distributed agent control architecture (Tele-MACS) describes how the coupling of planning agents with reactive agents can adapt the network resource configuration to comply with changes in user demand. The changes to the VPC route topology or capacity assignments are a requirement for the network to adapt to user behaviour and our control architecture demonstrates a mechanism for achieving the reconfigurations. The network scenario is considered a closed system, where the network operator has full control over the network elements, such as the routing tables and switching capabilities. We demonstrate our concepts with a focus on a core node ATM wide area network example, where the control system seeks to maintain network survivability and efficient use of bandwidth resources.

1 Introduction

Telecommunications management has traditionally been tackled via a tried and trusted centralised approach. For circuit switched type telecommunications networks the dynamics of the logical configuration was kept to a minimum with, for example fixed bandwidth reservations. With the advent of flexible platform technologies, (such as ATM - Asynchronous Transfer Mode) where dynamic run-time re-configuration of network resources allows for the realisation of specialised services (Weihmayer and Brandau 1990), the dynamics of the logical configuration are revolutionised.

Although ATM technology was conceived to satisfy these requirements one of the major problems is with the management of the resultant complexity. One of the approaches used to tackle this problem is to distribute the control and allow specialised agents to handle the localised tasks. Another approach for reducing complexity issues is to consider hierarchical control (Sommers 1996). The approach described exploits the strengths of each of these methods.

For this paper we define the agent software system for the control of logical network configuration, to be an intelligent control system (ICS). The proposed ICS described is intelligent due to its ability to plan tasks and deal with network contingencies. We do not require that an ICS deal optimally with all encountered exceptions, but it must act in a manner to manage the exception. The ICS that we have built consists of heterogeneous multiple collaborating agents to manage the logical configuration of a broadband telecommunications system. The agents are segregated into layers based on the two major mechanisms used by agent researchers (namely Deliberative and Reactive). To date there are few systems that have tackled the combination of these paradigms. Such agent systems are denoted as "hybrid frameworks" (Wooldridge and Jennings 1995), (Nwana 1996). Some of the arguments for the use of hybrid frameworks are quite intuitive, for example reaction is generally more robust, timely, and adaptive (Ferguson 1992). The planning attribute deals with more global, long term considerations of the system.

Generally planning systems have based their emphasis on the creation of a sequence of actions to be carried out such that a goal state is arrived at. One of the major problems with this kind of planning is that the qualitative or quantitative models make assumptions about the functioning of the system and often assume that events occur with certainty. Of course this is a fundamental problem in intelligent systems control. We feel that planning for system behaviour should be directly based on the modelling of the uncertainties that may occur in the functioning of the system. For example in this paper the agent control system maximises an expected profit related utility function that directly incorporates aspects of 'survivability' to uncertain event occurrences. This is quite distinct from developing a model of the system behaviour and incorporating uncertainty techniques to maximise long term profit. This is about planning using a model that explicitly incorporates a utility function that tries to minimise the probability of the contingencies damaging the system being controlled.

Considering the above arguments for the coupling of deliberation and reaction in control systems, we have developed a multi-level agent framework that contemplates the augmentation of planning and reactive heterogeneous (specialised) agents cooperating in a common world. The framework has been modelled on aspects of Brooks' Subsumption Architecture (Brooks 1986), but extends the concepts to consider multiple interacting agents that are in control of the world environment rather than alternatively operating in a world environment (e.g. as in Brook's Subsumption Architecture Mobile Robot).

One of the most difficult tasks to achieve in agent based systems is how to obtain co-ordination of the multiple agents so that advantageous work is achieved. One approach (taken by most researchers) is to tackle this problem by considering agent internal architectures, communication protocols, commitment schemes, scheduling algorithms and then couple them to generate a complete system (some would argue that this is the bottom up approach). This paper considers an alternative approach where we build the ICS on the basis of competence layers. That is, we build the *complete* co-ordinated system that achieves the *specified purpose* to a certain degree of competence (a top-down prototype approach). We then test the system and make adjustments until we are satisfied that the system layer meets its intended purpose. Next we build another complete layer of control to a *higher level* of competence and couple them through a suppression mechanism (see later). The key elements to note in this approach is that, layering allows us to isolate or encapsulate the interacting agents within a certain environment which aids extensibility (well defined interfaces between the layers are created) and that this provides *robustness* to software failure and robustness against the inability to reach a action sequence within the time interval of the control loop. The next section will describe some ATM telecommunications concepts. Following that the discussion will return to the layered agent control description.

2   Telecommunications - ATM Management

In ATM networks, a link fault tends to mean that a huge amount of flow is lost within a short period of time (Murakami and Kim. 1996), it is therefore fundamental that corrective action takes place extremely rapidly to minimise the loss. One of the competing goals in a network management system must therefore be the survivability of the network when faced with link or node failure. Most of the research into solving these problems use optimisation techniques, such as non-linear programming formulations. The solutions in the literature tend to be centrally controlled optimisation algorithms (Logothetis and Shioda 1995). Many papers do not directly focus on the ability to adapt network configuration to evolution of network or user behaviour. This paper proposes a distributed agent based control mechanism to deal with some of the intrinsic uncertainties in network management and its robustness will be demonstrated in a case study of VPC configuration management. In later sections 'survivability' metrics will be described, that provide a suitable utility function to balance the flow through the VPC network, where the maintenance of residual capacity provides enough resource to accommodate unexpected demand fluctuations caused by say surges in demand, link failures or node failures.

In ATM networks data is transported across the network by 53 byte cells following a Virtual Channel Connection (VCC) - For the context of this paper a VCC is set-up to provide a bi-directional connection, e.g. a voice call. A Virtual Path Connection (VPC) is a labeled route which can be used to transport, process and manage a bundle of VCCs. For the context of this paper, VPCs extend from a source node to a destination node and VCCs can be 'bundled' into these pre-assigned, but re-configurable VPCs. To make this description more concrete figure 1 (below) shows a simple network. The network consists of multiple demands occurring at time t1. Each demand encodes traffic intensity (cells per second), also the Class-of-Service (CoS) that the demand requires, and specifies the source-destination node pair. If we assume that the network has multiple VPCs between all source-destination node pairs already in operation at time t1, it is a control plane operation for the routing part of the CAC (Connection Admission Control) algorithm to decide the VPC that a connection request can take. Each source-destination VPC takes, in general, different physical node hops (again this is for maintaining physical survivability).

Fig. 1. Simple network topology and anticipated demand profile

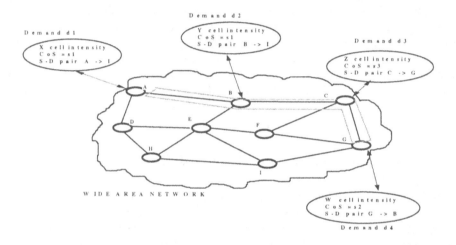

Figure 1 shows an ATM network topology example. The circles represent switching nodes and the thin black lines represent the physical links that interconnect them. It is possible to create end-to-end VPCs between source destination node pairs. For example, this is shown on the diagram as the 'pipe' structure following the path from node A to G via nodes B and C. We refer to this as a VPC. It is possible to set-up many VPCs through the network. Demand at the source nodes is converted into appropriate VCCs which are bundled into the VPC topology, as described earlier.

2.1   Virtual Path Connection Management

VPC management in ATM networks is usually considered as the management of two key concepts, VPC topology overlay (the route or sequence of nodes that the VPC traverses) and VPC capacity allocation (the bandwidth value assigned to the VPC). Although there is a large amount of research that addresses these two concepts, it is still unclear which method is best implemented. Each solution tends to use different cost metrics, such as minimising the call blocking probability, minimising hop counts, maximising call through-put, maximising survivability under fault conditions and trading control costs with set-up and efficiency gains. The extent to which VPC provisioning is able to improve network efficiency is highly dependent on its ability to provide VCCs with low set-up and switching costs, while maintaining a low call blocking probability. Traffic level fluctuations in networks generally cause a drop in network performance and hence require a response, either from routing or other mechanisms (Cuthbert and Sapanel 1993). Some of the factors that can cause changes in network performance are highlighted below:

- User groups changing behaviour - introduction of a new service.
- Periodic changes - hourly /daily/ monthly.
- Changes in the number of calls an individual makes.
- Change in popularity of a QoS type - tariff changes, new service introduction.

The ability to accommodate these changes is dependent on dimensioning of the VPC topology and capacity allocations resulting from resource management decisions (Friesen et al,1996). An alternative approach is to re-configure the VPC topology and the capacity allocations at run-time. This has been previously proposed in (Hayzelden and Bigham 1997a) and we build upon this by detailing some of the methods for achieving this.

3   Distributed Multi-Agent Control

Within complex distributed control domains, control is often achieved through the mix of algorithmic operations and subjective control from 'managers' of the control system. This kind of interaction occurs in ATM network management, where control operations such as Connection Admission Control (CAC) are invoked as an algorithm, whereas at present, human operators set-up ATM VPCs. As ATM becomes more widely used and Wide Area Network (WAN) providers adopt the technology, there will increasingly be a need for intelligent software control, to handle the complexities of allocating, routes (Frei and Faltings 1997), bandwidth and re-configuration of resource allocations (Kuwabara et al, 1996), (Wellman 1995) and also the dependency of the network technologies to make these factors possible (Tennenhouse et al, 1997), (Pitt and Mamdani 1997), (Hayzelden and Bigham 1998b).

Agent based management of telecoms networks has been researched by others including (Weihmayer and Brandau 1990), (Sommers 1996), (Miller et al, 1996). Sommers (Sommers 1996) previously focussed on a variety of aspects in ATM management including VPC management. However, this paper is more concerned with dynamic virtual path management specifically within the ATM reference plane and the re-configuration thereof. Appleby and Steward (Appleby and Steward 1992) explain their application of mobile agents to congestion control problems in circuit switched networks and in doing so, show that there is good reason for the requirement that there should be little or no direct inter-agent communication within complex control domains. One way of providing this in an agent based system is to provide layers of control. Our proposed multi-agent architecture similarly builds on a successfully implemented framework from the domain of robot control, namely Brooks' 'Subsumption Architecture' (Brooks 1986).

In the Subsumption Architecture, a Competence level is an informal specification of a desired class of behaviours. Each Competence is arranged in a hierarchical manner, where any level of Competence is dependent on the correct functioning of lower levels and completely independent of the higher Competence levels. The higher levels can suppress and or inhibit lower levels of behaviour, but this is generally the only communication between the layers that occurs (minimising communications overhead). We focus our attention on the advantages of the Subsumption Architecture, namely its robustness, reduced communications via layers and extensibility. However, we elaborate by integrating an element of planning behaviour at our higher management layers. Note that we are using concepts from the Subsumption Architecture as a framework for decomposing the agent activities and their relationships, not in a sense of an implementation model, i.e. each individual agent in our architecture is not based on the Subsumption Architecture. This is in accordance with research into hybrid control theory carried out by (Mataric 1992). A more detailed description of how the multi layer agent control architecture (figure 2) has been composed and how it relates to Subsumption concepts can be found in (Hayzelden and Bigham 1997b).

3.1   The Distributed Multi-layered Agent Control Architecture (Tele-MACS)

Figure 2, sketches the multi-layer multi-agent control architecture. We call this architecture Telecommunications Multi-Agent Control System (Tele-MACS). Firstly we will describe the basis of each of the principle agents and the tasks that they perform. Following this we will describe the details of the agent models and finally describe a particular topology reconfiguration example, including some of the results obtained from simulations.

A co-ordinating Tactical Survival Agent (TSA) holds a global VPC topology overlay model and also encodes the direct relationships to the underlying physical network infrastructure, such as the finite bandwidths of the individual links (generally optical fibres). The Tactical Survival Agent operates in the management plane of the ATM reference model and therefore has the required time to generate VPC topology plans such that predicted demand can be accommodated within the appropriate regions of the logical network structure. Once the plan for the next time interval is available the TSA makes available the new world state (VPC bandwidth values) to the control plane agents.

The control plane agents carry out operations considering the TSA emulated view of the world. Note that the TSA produces different views for the different control plane agents, such that, its plan action sequence is carried out to achieve the overall objective of implementing the planned VPC configuration, i.e. the TSA has the ability to suppress the inputs to each of the individual control plane agents. Note that the control plane agents are autonomous.

The TSA and a Connection QoS (Quality-of-Service agent) Agent operate at the ATM management plane time scale and so operate at a slower time scale. The Connection QoS Agent is extremely important to the maintenance of the network operator's service provision. The connection QoS Agent is only activated when conditions on the network mean that connections are beginning to be rejected, i.e. when the TSA cannot generate a plan to accommodate the predicted demand (there may not be enough physical resources available on the network or QoS constraints may become violated). The actual invocation of the QoS Agent would occur as resource saturation on certain VPCs became imminent.

The Charging Agent is implemented as a statistics gathering database based on the mechanism for segregating traffic types into attributed VPCs. Although the Charging Agent is not currently involved in the control mechanism, we hope to incorporate the charging information as a partial control metric. This will hopefully implicitly impose limits on the VPCs that a user connection requests. For example, applying a higher cost to certain high quality VPCs will reduce the likelihood of a user requesting that VPC.

Fig. 2. Telecommunications Multi-Agent Control System (Tele-MACS) Architecture

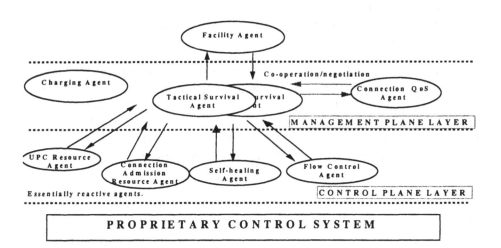

In the control plane there are multiple agents to control the routing part of ATM CAC (Connection Admission Control) for each source-destination node pair and a defined CoS. These Connection Admission Resource Agents (CARA) are essentially reactive agents that place connections (VCCs) into the available VPCs based on their default behaviour (they will be discussed later). There are also UPC (Usage Parameter

Control) Resource Agents, which allow improved throughput when the forecast context allows.

Although the TSA and connection QoS Agent have a global but abstract view of the network operational state, the control plane agents have the ability to carry out their tasks without the specific inter-communication of the management plane agents, albeit at a lower level of competence. The Self-Healing Agent will be a distributed implementation of the Self-Healing algorithm (Murakami and Kim 1996) that aids the generation of alternative paths under link failure conditions.

4   Principal Agent Control Actions

This section will describe the model used to generate the logical VPC topology plan. The mathematical model described below, shows how 'survivability' can provide a utility metric for creating VPC topologies that are robust to demand exceptions (uncertain events) and help prevent the network from becoming 'stressed' (operating at peak utilisation) (Meyer et al, 1995).

We are using Quality of Service (QoS) as the most important utility factor, i.e. that maximising traffic throughput in high speed networks is not necessarily the most valued parameter. In fact from a network operators perspective maintaining QoS means that customers are more likely to have their service requirements met and therefore be satisfied with the service provision.

In this description we suppose there is only one CoS for each source destination pair. (The procedure for planning feasible desired changes in bandwidth allocation can be generalised to many classes of service). A source-destination pair is represented by the variable[1]. Let $VP(\pi)$ be the set of all acyclic end to end VPC paths for this source-destination pair. The flow (in cells per second say) through a particular path $s \in VP(\pi)$ is denoted by f_s^{π}. A path s for source-destination pair π is denoted by Q_s^{π}. We will imagine that $f_s^{\pi}(t_0)$ are observable.

The demand at time t_0 for a source-destination pair is $q^{\pi}(t_0)$ where

$$q^{\pi}(t_0) = \sum_{s \in P(\pi)} f_s^{\pi}(t_0)$$

The capacity of link i, v_i, for all links in the network is also assumed as known.

We assume that we have some forecasting mechanism which provides forecasts of $q^{\pi}(t)$ for t' > t_0. The forecast demand for time t' given the actual demand $q^{\pi}(t)$ at

---

[1] The problem can be formulated in terms of arcs rather than paths, but with our assumption of end to end VPCs a path formulation is more natural at this stage.

time t is denoted by $Q^{\pi}(t'|q^{\pi}(t))$. This can use a mixture of historical profiles and tracking.

Using the computing resource bound allocated, the TSA's planning mechanism looks ahead to establish optimal feasible adjustments to $f_s^{\pi}(t_0)$ so to achieve a desired $f_s^{\pi}(t')$ where t' is a time ahead greater than the control cycle time. The optimal solution computed, allows for the capacity constraints and the inertia of the system being controlled. (For example suppressing connections along a path will not result in an instantaneous drop in flow and so there is no point in generating a solution, which demands disconnection rates which are not likely. The fall in link usage depends on the type of call and their typical connection time). Once the desired changes in flow are computed they are scaled down pro rata to the duration of the next control cycle.

To guide the flow adjustment two loss functions are being experimented with. They are *minimising the maximum residual in the links* and *minimising the expected cell loss on link failure*. The former is simple and has the desirable property that if demand increases proportionally at each source then the solution remains optimal. The latter is based on emulating the manner in which cells can be lost in an ATM network on link failure. Here when a link fails, connections on that link are re-routed along the shortest alternative path for this link, any remaining are routed through the link disjoint second shortest path, etc. Any cells, which cannot be re-routed are lost.

The total flow in link i is $x_i = \sum\limits_{\pi \in SD} \sum\limits_{s \in VP(\pi)} P_s^{\pi}(i) f_s^{\pi}$, where SD is the set of all source destination pairs. $P_s^{\pi}(i)$ is 1 if link i is on path Q_s^{π}, and 0 otherwise. The capacity of link i is denoted by v_i and the residual capacity by $y_i = v_i - x_i$.

So for example, the expected loss for one alternative path is used,

$$\sum\limits_{i=1}^{L}(x_i - \min\limits_{j \in SP(j)}\{y_j\})p_i$$ where SP(j) is the shortest alternative path for link j, p_i

is the probability that link i fails.

Maximising the minimum residual corresponds to $\max\limits_{f_s^{\pi}} \min\limits_{i \in L}\{y_i\}$, where L is the set of links.

4.2 Simple Network Survivability Example

Example: Take a simple network with 6 source destination pairs, AB, BA, BC, CB, AC, CA, each having a demand at t_0 of 100 (i.e. $q^{ab}(t_0)=100$, $q^{ba}(t_0)=100$, etc) and each with two paths. VP(AB) is {AzyB, AxB} and each has a flow of 50, i.e. $J_{AzyB} = 50$ and $J_{AxB} = 50$. The other source destination pairs have similar and symmetrical VP flows.

Fig. 3. The initial capacities and demands are shown.

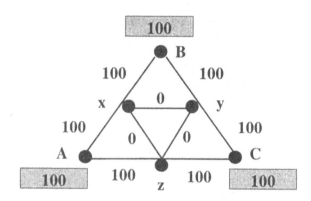

Suppose the forecast demand for each source-destination pair is 120 then the computed optimal flow under both criteria leads to residual capacities shown in figure 4. (Links are assumed to be equally likely to fail)

Fig. 4. The final capacities and demands are shown.

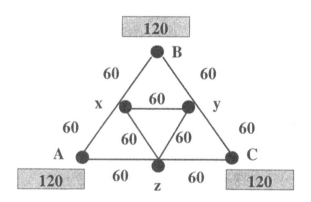

So the changes in flow for s-d pair 1 is: $D f_{AxB}^{AB} = 50$, $D f_{AzyB}^{AB} = -30$. Similarly for the other s-d pairs. Further details of the metrics and mathematics that the survivability utility function is based upon can be found in (Bigham and Hayzelden 1998).

4.3 Control of the *Routing* Part of Connection Admission Control

The above sections have described the Tactical Survival Agent model, such that a VPC topology plan can be created. For this plan to be implemented successfully the control plane agents must be co-ordinated appropriately. This section describes how the control plane agents are influenced by the management plane agents so that the VPC topology can be re-configured in accordance with the plan.

There has been a large amount of research into CAC algorithms and it is not our intention to redo this. Our current strategy is to use whatever CAC algorithm is provided and use the information from our management plane agents to suppress Connection Admission Resource Agent (CARA) inputs (Brooks 1986).

For example, consider a change from one VPC bandwidth configuration to the next planned configuration. Say we have three VPCs between nodes A and B (see figure 5). The Tactical Survival Agent generates a plan for the next time period. The plan may suggest that for the predicted demand profile, the required actions are to increase the bandwidth in VPC 1 and that the other two VPCs must be reduced by a calculated summed proportion. Since the Tactical Survival Agent has the 'knowledge' that 'soon', the capacity of VPC 1 will increase to its new specified capacity (to cope with predicted demand), it would be extremely advantageous to aid the change over to the new configuration, by giving the CARA some guidance to which routes are the most appropriate for use at the actual time interval. We are therefore guiding the VPC route selection process based on forward time interval knowledge.

The mechanism chosen for guiding the control plane agents is via suppression. For example suppressing the CARA's input so that the agent chooses, in priority to the others, the VPC that will have its bandwidth increased in the immediate future (there will be more resources available on this VPC to accommodate the connection). Supposing that the TSA wants the capacity of VPC 2 and 3 to decrease (as shown in figure 5 by dotted lines), it does not want any further connections to be placed within these VPCs. It therefore 'tricks' (via suppression) the CARA into believing that VPC 2 and 3 do not exist. This is achieved by providing the CARA with a suppressed view of the world. The CARA then autonomously assesses whether VPC 1 can accept the connection request (the only remaining VPC according to the CARA's suppressed view), by invoking the CAC multiplexing check algorithm. CAC multiplexing check algorithms decides whether enough bandwidth resource remains on the VPC for the connection to be accepted (Pitts and Schormans 1996). In the mean time the connections within VPC2 and VPC3 will be reducing at a rate determined by the mean call duration (connections will be terminating), which is directly dependent on the CoS for VPCs in question[2]. In this way the bandwidths of these VPCs can be re-calibrated in compliance to the Tactical Survival Agent's next plan.

---

[2] For example the average time duration for voice type connections may be, say five minutes.

After some time, the suppression can stop and the new updated view of the logical network configuration can be supplied to the CARA so that it can continue to allocate connections to the VPCs. The control mechanisms that we have explained also allows for the possibility for adapting the VPC route topology, rather than just re-allocating bandwidths (Hayzelden and Bigham 1997a), although this will not be discussed here.

Figure 5 below, schematically depicts the scenario described, where A and B represent the source destination nodes respectively. The widths of each VPC diagram relate to the relative bandwidths. The solid lines show the bandwidth state at present, and the dotted lines represent the planned bandwidths required for the next time interval, i.e. VPC 1's bandwidth will increase in the next time interval.

Fig. 5. Agent operating at a S-D pair for the associated VPC

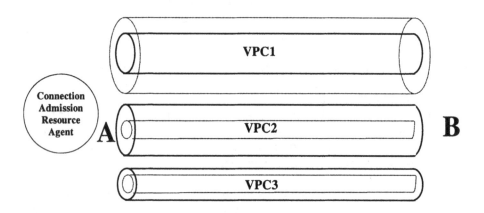

4.4 High Level Facility Agent Activity and Tasks

The Facility Agent has a very high level model of the system and is only invoked when the Tactical Survival Agent passes a parameter suggesting that an update of the physical topology is required to cope with future demands, i.e. when the VPC configuration is no longer able to adapt to the changes in demand due to the finite limits of the physical links.

5   Preliminary Results

The work is currently being validated through the use of an ATM connection level discrete event simulation implemented in C++ and Java.

Results showing the effectiveness of suppressing inputs to the CARA are shown below. In these preliminary simulations we have created three VPCs for a source-destination pair and a particular CoS, as in figure 5. The simulation generates connections over a simulated period of one hour. The CARA allocates the connection requests to the pre-assigned VPCs by creating VCCs (as described in section 4.3). A simple CAC algorithm carries out a bandwidth multiplexing check on the chosen VPC and if it returns true, the connection (VCC) is established.

The graph shows the amount of weighted bandwidth within the three VPCs available after the simulation has completed. The two implemented algorithms shown are: Round-Robin connection allocation (dotted line) and Tactical Survival Agent suppression allocation (solid dark line).

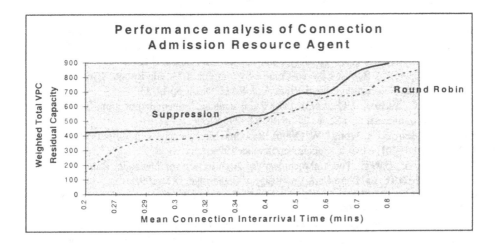

Fig. 6. Agent based Suppression mechanism (solid line) vs. Round Robin algorithm (dotted)

The results show that the control scheme based on the survivability model (suppression from the management plane agents) provides the VPCs with a higher relative proportion of bandwidth availability at the end of the simulation period - this should therefore aid the logical network re-configuration that maybe necessary in the next time frame to achieve better utilisation of resources and reduction of connection blocking. Further experiments will investigate the connection rejection rates over multiple time intervals and therefore provide a suitable performance measure.

Acknowledgements

This research is partly funded by EPSRC and British Telecom (BT Labs).

6 Conclusion

This paper has described a multi-layered multi-agent architecture for the management of virtual path configuration in ATM networks. The coupling of planning agents (management plane layer) operating at a slower time scale to a reactive agent layer (control plane) via a simple suppression communication, provides a novel mechanism to achieve logical network topology re-configuration that is theoretically robust and extensible.

7 References

Appleby S., Steward, S. (1994) 'Mobile software agents for control in telecommunications networks.' BT Technology Journal, Vol.12. No. 2. April 1994. pp.104-113.

Bigham, J., Hayzelden, A. (1998). Using Survivability to Manage Uncertainty in a Telecommunications Network Management Application, Technical Report, Dept. of Electronic Engineering, Queen Mary College, Univ. of London, U.K.

Brooks, R (1986) 'A Robust Layered Control System For A Mobile Robot.' IEEE Journal of Robotics and Automation. March 1986. Vol. RA-2. No.1. pp.14-23.

Cuthbert, L, G., Sapanel, J, C. (1993)'ATM The Broadband Telecommunications Solution' IEE Telecommunications Series 29. ISBN 0 85296 815 9.

Friesen V J., Harms J J., Wong J W. (1996) 'Resource Management with Virtual Paths in ATM Networks' IEEE Network, September/October 1996. pp. 10-20.

Ferguson, I., A., (1992) 'TouringMachines: An Architecture for Dynamic, Rational, Mobile Agents', PhD Thesis, Clare Hall, University of Cambridge, U.K. 1992.

Frei, C., Faltings, B., (1997) 'Intelligent Agents for Network Management'. AI for Network Management Systems 1997, IEE Digest No. 97/094.

Hayzelden, A. & Bigham, J. (1997a) 'Management of ATM Resource Configuration, Automated by Distributed Agent Control.' Proceedings of the IEE Fourth Communication Networks Symposium, Manchester 7- 8 July 1997.

Hayzelden, A. & Bigham, J. (1997b) 'Subsumption Approach Applied to the Decomposition of Multi-Agent Software Systems and Telecommunications Management' Technical Report, Department of Electronic Engineering, Queen Mary College, University of London. 1997.

Hayzelden, A. & Bigham, J. (1998a) 'Software Agents for Telecommunications Management and Control' In Proceedings of IEEE Communication Systems and Digital Signal Processing. Sheffield, U.K.

Hayzelden, A (1998b) 'Telecommunications Multi-Agent Control System (Tele-MACS).' To appear in proceedings of ECAI'98, Brighton, U.K.

Kuwabara, K. Ishida, T. Nishibe, Y. Suda, T (1996) 'An Equilibratory Market-Based Approach for Distributed Resource Allocation and its applications to Communication Network Control.' Clearwater, S, H. Market-Based Control: A Paradigm for Distributed Resource Allocation, World Scientific, 1996 Chapter 3.

Logothetis, M., Shioda, S. (1995) 'Medium-Term Centralized Virtual-Path Bandwidth Control Based on Traffic Measurements.' IEEE Transactions on Communications, Vol. 43, No. 10, October 1995. pp.2630-2640.

Mataric, M J. (1992) 'Behavior-Based Control: Main Properties and Implications.' Proceedings, IEEE International Conference on Robotics and Automation. Workshop on Architectures for Intelligent Control Systems, Nice, France, May 1992, pp 46-54

Meyer, K., Erlinger, M., Betser, J., Sunshine, C., Goldscmidt, G., Yemini, Y. (1995) 'Decentralising Control and Intelligence in Network Management.' Integrated Network Management IV, Chapman & Hall, 1995. Ed. Sethi et al. pp. 4-15. ISBN 0412715708.

Miller, M, S., Krieger, D., Hardy, N., Hibbert, C., Tribble, E, D. (1996) 'An automated auction in ATM network bandwidth'. In Clearwater, S, H. Market-Based Control: A Paradigm for Distributed Resource Allocation, World Scientific, 1996 Chapter 5.

Murakami K., Kim, H S. (1996) 'Virtual Path Routing for Survivable ATM Networks.' IEEE/ACM Transactions on Networking Vol.4 No.1 February 1996. pp. 22-39.

Nwana, H S. (1996) 'Software agents: an overview.' The Knowledge Engineering Review. 1996. Vol.11. No.3. pp.205-244.

Pitt, V, J., Mamdani, E, H., Hadinham, R, G., Tunnicliffe, A, J. (1997) 'Agent-Oriented Middleware for Telecommunications Network & Service Management' AI for Network Management Systems 1997, IEE Digest No. 97/094

Pitts J, M., & Schormans, J., A. (1996) 'Introduction to ATM Design and Performance.' John Wiley & Sons Ltd, 1996, pages 93-112 - ISBN 0-471-96340-2.

Somers, F. (1996) 'HYBRID: Intelligent Agents for Distributed ATM Network Management' In IATA'96 Workshop at ECAI'96, Budapest, Hungary, 1996.

Tennenhouse, D., Smith, J., Sincoskie, W, D., Wetherall, D. (1997) 'A Survey of Active Network Research.' IEEE Communications Magazine, 1997.

Weihmayer R, Brandau R. (1990) 'Cooperative distributed problem solving for communication network management' Computer Communications, Vol.13, No. 9. November 1990. pp.547-556.

Wellman, M. (1993) 'A Market-Oriented Programming Environment and its Application to Distributed Multicommodity Flow Problems' Journal of Artificial Intelligence Research 1 (1993) pp. 1-23.

Wooldridge, M and Jennings, N. (1995) 'Intelligent Agents: Theory and Practice'. The Knowledge Engineering Review. Vol.10. No. 2. 1995. pp.115-152.

Routing in Telecommunications Networks with Ant-Like Agents

Eric Bonabeau[1], Florian Henaux[2], Sylvain Guérin[3], Dominique Snyers[3], Pascale Kuntz[3], Guy Theraulaz[4]

[1] Santa Fe Institute, 1399 Hyde Park Road, Santa Fe, NM 87501, USA
bonabeau@santafe.edu
[2] Ecole Nationale Supérieure des Télécommunications de Paris, 46 rue Barrault, 75634 Paris Cédex 13, France
[3] Ecole Nationale Supérieure des Télécommunications de Bretagne, BP 832, 29285 Brest Cédex, France
[4] CNRS - UMR 5550, Laboratoire d'éthologie et psychologie animale, Université Paul Sabatier, 118 route de Narbonne, 31062 Toulouse, France

Abstract. . A simple mechanism is presented, based on ant-like agents, for routing and load balancing in telecommunications networks, following the initial works of Appleby and Stewart [1] and Schoonderwoerd et al. [32,33]. In the present work, agents are very similar to those proposed by Schoonderwoerd et al. [32,33], but are supplemented with the ability to perform more computations at switching nodes, which significantly improves the network's relaxation and its response to perturbations.

1 Introduction

1.1 Routing in Telecommunications Networks

Routing is a mechanism that allows calls to be transmitted from a source to a destination through a sequence of intermediate switching stations or nodes, because not all points are directly connected: the cost of completely connecting a network becomes prohibitive for more than a few nodes. Routing selects routes that meet the objectives and constraints set by the user traffic and the network, and therefore determines which network resources are traversed by which user traffic [26,34]. Why should routing be (1) dynamic, and (2) decentralized?

(1) The pathway of a message must be as short as possible. One solution is therefore to design fixed routing tables such that any two nodes in the network are connected through the shortest possible path. Designing such routing tables is a simple optimization problem, which has to be solved when the network topology has been defined. But traffic conditions are constantly changing, and also the structure of the network itself may fluctuate (switching stations or connections can fail). The problems of telecommunications networks is to minimize the number of call failures in any condition. Because there are usually many possible pathways for one message to go from a given node to another node, it is possible in principle to make routing algorithms adaptive enough to overcome local congestions: calls can be rerouted to nodes that are less congested, or have spare capacity. If there is a sudden burst of activity, or if one node becomes the destination or the emitter of a large number of calls, rerouting becomes crucial. Static routing, whereby routing remains fixed independent of the current states of the network and user traffic, is therefore almost never implemented: most routing schemes respond in some way to changes in network or user traffic state. But there exists a wide spectrum of dynamic routing systems, which vary dramatically in their responsiveness, speed of response and in the types of changes they respond to [34]: some routing systems can be seen as "quasistatic", because routing is modified only in response to exceptional events (link or switch failure), and/or on a long time scale, while other routing systems are highly dynamic and autonomously update traffic routing in real time in response to perceived changes in user and network state. Tightly coupled with dynamic routing is load balancing, which is the construction of call-routing schemes that successfully distribute the changing load over the system and minimize lost calls. Load balancing makes it possible to relieve actual or potential local congestion by routing calls via parts of the network that have spare capacity. Dynamic routing and load balancing require more computational resources than static or quasistatic routing. It relies on active participation of by entities within the network to measure user traffic, network state and performance, and to compute routes.

(2) Most routing algorithms today are centralized, with routing tables at switching stations being updated by a central controller at regular intervals. But:

• The controller needs current knowledge about the entire system, necessitating communications links from every part of the system to the controller.

• Central control mechanisms scale badly.

• Failure of the controller leads to failure of the whole system.

• Telecommunications networks are distributed, extended, dynamic, highly unpredictable systems, and central control may simply not be appropriate.

• A routing system's responsiveness to state changes depends upon the load on the central controller and the distance between the central controller and the portion of the network requiring adaptation.

• Centrally controlled systems might have to be owned by a single authority.

• By contrast, with a decentralized implementation of the routing function, multiple entities function independently and exchange information, providing fault tolerance. Moreover, the routing computational load is spread among the multiple entities.

1.2 Agent-Based Routing and Network Control

The idea that ant-like agents, or "mobile software agents", could be used for network control in telecommunications has been introduced by Appleby and Steward [1,22], in a paper that poses the problems clearly but remains vague as regards actual implementation. Schoonderwoerd et al. [32,33] have proposed an interesting version of Appleby and Steward's [1] work, where they use simple agents that modify the routing tables of every node in the network. Their work, being described in more detail, is somewhat easier to reproduce, and contains a set of nice ideas. Routing and load balancing mechanisms based on their algorithm may conceivably be used in some networks in the near future, if it can be proved, one way or another, that *any* network condition can be satisfactorily dealt with. New approaches to routing require intensive testing.

1.3 Ant-Like Agents

Schoonderwoerd et al.'s [32,33] work relies on the ability of social insects to solve problems, sometimes difficult problems, in a distributed way, without any central control, on the basis on local information. Social insects also often exhibit flexibility (they can respond to internal perturbations and external challenges) and robustness (failure of one or several individuals usually does not jeopardize a colony's functioning). Given these properties and the impressive ecological success of social insects [38], it does not seem unreasonable to try to transfer current knowledge about how insect societies function into the context of engineering and distributed artificial intelligence. This approach is similar to other approaches consisting in imitating the way "nature" (that is, physical or biological systems) solves problems [7,19]. Another possible and apparently very promising pathway is to use economic or financial metaphors to solve problems [24,25].

Recent research in ethology suggests that self-organization is a major component of a wide range of collective phenomena in social insects [15,5]. Theories of self-organization, originally developed in the context of physics and chemistry to describe the emergence of macroscopic patterns out of processes and interactions defined at the microscopic level, can be extended to social insects to show that complex collective behavior may emerge from interactions among individuals that exhibit simple behavior. A good illustration of the self-organized, distributed problem-solving ability of social insects is the binary bridge experiment [2,15]: in experiments with the ant *Linepithema humile*, a food source is separated from the nest by a bridge with two branches A and B, with branch B being r times longer than

Figure 1a. Experimental setup and drawings of the selection of the short branches by a colony of *L. humile* respectively 4 and 8 minutes after the bridge was placed.

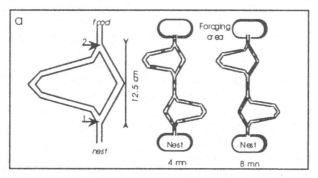

branch A (Fig. 1a). It is observed that in most experiments, the short branch is selected by the colony if r is sufficiently large (r=2 in Fig. 1b). This is because these ants have a trail-laying/trail-following behavior: individual ants lay a chemical substance, a pheromone, which attracts other ants. The first ants returning to the nest from the food source take the shorter path twice (from the nest to the source and back), and therefore influence outgoing ants towards the short branch which is initially more strongly marked. If, however, the short branch is presented to the colony after the long branch, the short path will not be selected because the long branch is already marked with pheromone: the colony does not exhibit flexibility (Fig. 1b). This problem can be overcome in an artificial system, by introducing the pheromone's lifetime: because pheromone diffuses and evaporates, it is more difficult to maintain a stable pheromone trail on a long path than on a short path. In

Figure 1b. Distribution of the percentage of ants that selected the shorter branch over all experiments (14 and 18 experiments, respectively). The longer branch is r times longer than the short branch. The second graph (n=18, r=2) corresponds to an experiment in which the short branch is presented to the colony 30 minutes after the long branch: the short branch is not selected, and the colony remains trapped on the long branch. From Ref. 15.

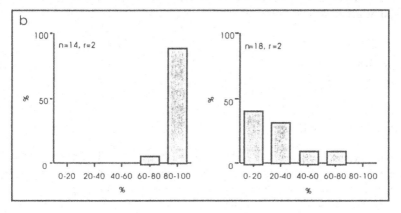

this way, the short branch can be selected even if presented after the long branch. In the real world, although pheromone concentrations do decay, the lifetime of pheromones is usually so large that it cannot allow such a flexibility: this is a case where there is a clear divergence from biological reality, but solving problems and understanding nature are two activities which have their own criteria of success.

Taking advantage of this ant-based optimizing principle, Colorni et al. [10,11,12], Dorigo et al. [17], Dorigo and Gambardella [18,20] and Gambardella et al. [21] have proposed an ingenious optimization method, the Ant System (or the Ant Colony System, which is more efficient), which they applied to classical optimization problems, such as the traveling salesman problem [30], the quadratic assignment problem [28] or the job-shop scheduling problem, with reasonable success: this method, as a general heuristic, can be compared to simulated annealing [27]. Others have followed the path, and have either extended the original method [8,35] or applied ant-based optimization to the vehicle routing problem [9], the graph coloring problem [13] and the search of continuous spaces [3,39]. Schoonderwoerd et al.'s [32,33] application to routing and load balancing in telecommunications networks is derived from the same metaphor: however, the formulation of their algorithm, the dynamic properties of the problem and the highly distributed nature of the underlying system, make this application original and of special interest. In particular, it is, until now, the only example where the advantage of using the decentralized functioning of social insect colonies is (and could be even more) fully exploited.

Other applications based on the functioning of social insect societies include data analysis [31] and graph clustering [29] inspired by cemetery organization in ants [16], adaptive task allocation [6] inspired by the division of labor in social insects [4,36], and self-assembly inspired by collective building in wasps [37].

1.4   Outline of the Paper

The rest of the paper is dedicated to describing and extending the ant-like-agent-based mechanism introduced by Schoonderwoerd et al. [32,33] to include a form a dynamic programming which makes the agent "smarter".

2   Algorithms

2.1   Original Algorithm [32,33]

If there are n nodes in the network, a node N_i with k(i) neighbors is characterized by a routing table $R_i = [r_{i,m}]_{n-1,k(i)}$ that has n-1 rows and k columns: each row corresponds to a destination node and each column to the next node. $r_{i,m}$ gives the probability that a given message, the destination of which is node N_i, be routed from node N_i to node N_m.

Agents update routing tables of nodes viewing their node of origin as a destination node: links being bidirectional, agents that have a certain "knowledge" about some portion of the network (where they come from) modify routing tables of nodes that influence the routing of messages and agents that have this portion of the network as destination. This avoids agents going back all the way to their node of origin to update all intermediate routing tables, and therefore decreases in pinciple the network's agent load.

Agents can be launched from any node in the network at any time. Destination is selected randomly among all other nodes in the network. The probability of launching an agent per unit time must be tuned to maximize the performance of the system. It appears that too few agents are not enough to reach good solutions, whereas too large a number of agents degrades the performance of the system by adding noise (a more appropriate way of launching agents would be to generate an increasing number of agents as the network becomes more and more congested). The notion of performance of the system here is simply the number of calls that fail to reach their destinations. Agents go from their source node to their destination node by moving from node to node. The next node an agent will move to is selected according to the routing table of its current node. Each visited node's routing table is updated: more precisely, an agent modifies the row corresponding to its source node, which is viewed as a destination node. When the agent reaches its destination node, it updates the local routing table and is deleted from the network. How are routing tables updated? Let N_s be the source node of an agent, N_m the node it just comes from, and N_i its current node at time t. The entry $r_{s,m}^i(t)$ is reinforced while other entries $r_{s,l}^i(t)$ in the same row decay:

$$r_{s,m}^i(t+1) = \frac{r_{s,m}^i(t) + \delta r}{1 + \delta r} \quad , \tag{1}$$

$$r_{s,l}^i(t+1) = \frac{r_{s,l}^i(t)}{1 + \delta r} , \tag{2}$$

where δr is a reinforcement parameter, that depends on the agent's characteristics. Note that this updating procedure conserves the normalization of $r_{s,l}$,

$$\sum_l r_{s,l}^i = 1 \tag{3}$$

if they are initially normalized, so that $r_{s,l}^i$ can always be considered as probabilities, all of which always remain strictly positive. $r_{s,m}^i(t)$ is comparatively more reinforced when it is small (that is, when node m is not on the preferred route to node s) than when it is large (that is, when node m is on the preferred route to node s). This is an interesting feature, as it should allow new routes to be discovered quickly when the preferred route gets congested because established routing solutions become more easily unstable: there is an exploration/exploitation tradeoff to be found, for too much instability might not always be desirable (one way of getting around this problem would be, once again, to increase the number of agents only when the network is congested, because more agents destabilize solutions that use congested routes).

The influence δr of a given agent must depend on how well this agent is performing. Aging can be used to modulate δr: if an agent has been waiting a long time along its route to its destination node, it means that the nodes it has visited and links it has used are congested, so that δr should decrease with the agent's age (measured in time units spent in the network). Aging should in principle be relative to the length (expressed in time units) of the optimal path from an agent's source node to its destination. Schoonderwoerd et al. [32,33] choose to use absolute age T, measured in time units spent in the network, and propose

$$\delta r = \frac{a}{T} + b,$$ (4)

where a and b are parameters. Aging is also used in a complementary way: age is manipulated by delaying agents at congested nodes. Delays result in two effects: (1) the flow of ants from a congested node to its neighbors is temporarily reduced, so that entries of the routing tables of neighbors that lead to the congested cannot be reinforced, and (2) the age of delayed agents increases by definition, so that delayed ants have less influence on the routing tables of the nodes they reach, if δr decreases with age. Schoonderwoerd et al. [32,33] suggest that the delay D imposed on an agent reaching a node with spare capacity S (that is, the percentage of slots left in the node for new messages) should be given by

$$D = c\, e^{-gs},$$ (5)

where c is a parameter and S_c is a characteristic spare capacity (expressed in percentage of the node's capacity), so that a node is considered congested if $S \gg S_c$.

Finally, one needs to avoid freezing routes in situations that remain static for a long time and then suddenly change. Finding new routes is facilitated by increased reinforcement of small entries, but this may be insufficient. Schoonderwoerd et al. [32,33] suggest the addition of a tunable "noise" or "exploration" factor f ($0 < f < 1$). At every node, an agent chooses a purely random path with probability f and chooses its path according to the node's routing table with probability ($1-f$). Noise allows to maintain information about apparently useless routes to give a head start when the preferred route is blocked. It also allows to rediscover quickly a better route that appears owing to the release from congestion of a node.

2.2 "Smarter" Agents

At each time step, an agent as described by Schoonderwoerd et al. [32,33] only updates the row that corresponds to its source node (viewed as a destination node). An interesting recent addition by Guérin [23] introduces updating of all rows corresponding to all the intermediate nodes visited by the agent, in a way reminiscent of dynamic progamming: reinforcement of an entry associated with a given node is discounted by a factor that depends on the agent's age relative to the time it visited that node. Although Guérin's [23] method relies on agents that perform round trips from their source node to their destination node and back, it can be readily applied to the case of one-way agents. In order to distinguish these more complex or "smarter" agents from those of Schoonderwoerd et al. [32,33], we call them "smart" agents. Let N_i be the ith node visited by an agent on its way to its destination. The agent updates the rows of all intermediate nodes in N_i's routing table. Updating of the row corresponding to node N_m ($m<i$), the mth visited node, is performed using a relative age instead of its absolute age. Entry $r_{m,i-1}(t)$ is reinforced while other entries $r_{m,l}(t)$ ($l \bullet i$) in the same row decay:

$$r_{m,i-1}^i(t+1) = \frac{r_{m,i-1}^i(t) + \delta r}{1 + \delta r},$$ (7)

$$r_{m,l}^i(t+1) = \frac{r_{m,l}^i(t)}{1 + \delta r},$$ (8)

$$\delta r = \frac{a}{T_i - T_m} + b,$$ (9)

where T_m is the agent's absolute age when reaching node N_m and T_i its age when reaching node N_i. The entries corresponding to nodes visited long ago are weakly updated, which is expressed in equation (9).

The ant-based algorithm based on smarter agents yields significantly better performance. Figure 2a shows the average number of call failures per 500 steps together with error bars representing the standard deviation observed over 10 trials of the simulation with identical parameters, the first 500 steps being discarded, after the network has been initialized. The parameters are identical to those of Schoonderwoerd et al. [32,33]: we use the 30 node interconnection network of British Telecom as an example; each node has a capacity of 40 calls (S=100%); during every time step, an average of one call is generated, with an average duration of 170 time steps; the probabilities that nodes be emitters or receivers are uniform in [0.01,0.07] and normalized; generation and normalization of emission and reception probabilities for all nodes defines a set of call probabilities, and a change in call probabilities means that new emission and reception probabilities have been generated (new nodes become more likely to be emitters or receivers, and others less likely). At initialization, routing tables are characterized by equiprobable routing (all neighboring nodes are equally likely to be selected as next node), and there is no message in the network. Messages are routed independently of the agents' dynamics: when a message reaches a node, it selects the largest entry in the appropriate row it its current table and is routed towards the neighboring node corresponding to this largest entry. During the first phase following initialization (from t=501 to 3000), smart agents perform significantly better than Schoonderwoerd et al.'s (1997) agents (t-test, df=18, t=3.2, P<0.003); during the second phase (t=3001 to 7500), when a stationary state in the dynamics of call failures has been reached, smart agents also perform significantly better (df=18, t=3.9, P<0.001), with a level of significance comparable to the one obtained during the first phase.

Figure 2a. Adaptation (500 first steps discarded). Phase 1: from t=501 to 3000, Phase 2: from t=3001 to 7500. a=0.08, b=0.005, c=80, S_c=13.3%, f=0.05, node capacity=40 calls.

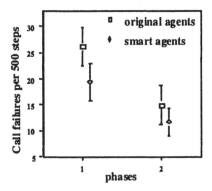

Figure 2b. Change in call probabilities (500 first steps discarded). Phase 1: from t=501 to 3000, Phase 2: from t=3001 to 7500. a=0.08, b=0.005, c=80, S_c=13.3%, f=0.05, node capacity=40 calls.

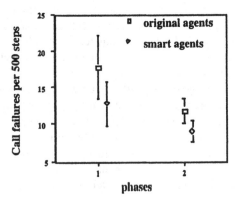

Figure 2b shows the average number of call failures per 500 steps together with error bars representing the standard deviation observed over 10 trials of the simulation with identical parameters, the first 500 steps being discarded, after a change in the call probabilities. Results similar to those obtained after network initialization are observed: during both phases, smart agents perform significantly better than simple agents (phase 1: df=18, t=4.1, P<0.001; phase 2: df=18, t=2.3, P<0.025), but the level of significance is much lower during the second phase, indicating that smart agents are particularly useful when network conditions are changing.

3 Limitations and Future Work

We have supplemented the agents introduced by Schoonderwoerd et al. [32,33] with a simple extension initially suggested by Guérin [23] with more complex agents. Analysis of call failures indicates that this addition improves significantly the performance of the routing scheme, especially when network traffic is subject to variations. But before this method can actually be implemented in real communications networks, some limitations have to be overcome. First, the model network used in this paper is obviously an oversimplification of reality: the method has to be tested on more realistic, and therefore more complex, network models. But this points to the problem of analyzing the routing's behavior. Routing algorithms are generally difficult to analyze mathematically, especially when the underlying network is complex and/or not fully connected: for example, the properties of the dynamic alternative routing method [26] could be obtained analytically only for fully connected networks. It is crucial for "self-organizing" algorithms, such as the one presented in this paper, where control by humans can only be limited, to be able to prove that they are not going to fall apart in some, possibly pathological but still potential, specific configuration. For example, it would be good to be sure that messages cannot become trapped in infinite cycles without ever reaching their destination. One also needs to have a clear understanding of the limits and constraints of communications networks: for example, if there are sufficient computational power and spare capacity in the network to launch a large number of complex agents without affecting traffic, why bother to design simple agents?

Acknowledgments

E. B. is supported by the Interval Research Fellowship at the Santa Fe Institute. E. B., F. H. and G. T acknowledge support from the GIS (Groupe d'Intérêt Scientifique) Sciences de la Cognition.

4 References

1. Appleby, S. & Steward, S. (1994). Mobile software agents for control in telecommunications networks. *British Telecom Technol. Journal* **12**, 104-113.
2. Beckers, R., Deneubourg, J.-L., Goss, S. and Pasteels, J.-M. (1990). Collective decision making through food recruitment. *Ins. Soc.*, **37**, 258-267.
3. Bilchev, G. & Parmee, I. C. (1995). The ant colony metaphor for searching continuous design spaces. In: *Lecture Notes in Computer Science* (Fogarty, Y., ed.) 993, 25-39, Springer-Verlag.
4. Bonabeau, E., Theraulaz, G. & Deneubourg, J.-L. (1996). Quantitative study of the fixed threshold model for the regulation of division of labour in insect societies. *Proc. Roy. Soc. London B* **263**, 1565-1569.
5. Bonabeau, E., Theraulaz, G., Deneubourg, J.-L., Aron, S. & Camazine, S. (1997a). Self-organization in social insects. *Trends in Ecol. Evol.* **12**, 188-193.
6. Bonabeau, E., Sobkowski, A., Theraulaz, G., Deneubourg, J.-L. (1997b). Adaptive task allocation inspired by a model of division of labor in social insects. In *Bio-Computation and Emergent Computing*, eds. D. Lundh, B. Olsson & A. Narayanan. Singapore: World Scientific.
7. Bounds, D.G. (1987). New optimization methods from physics and biology. *Nature* **329**, 215-219.
8. Bullnheimer, B., Hartl, R. F. & Strauss, C. (1997a). A new rank based version of the ant system: a computational study. Working paper #1, SFB Adaptive Information Systems and Modelling in Economics and Management Science, Vienna.
9. Bullnheimer, B., Hartl, R. F. & Strauss, C. (1997b). Applying the ant system to the vehicle routing problem. *2nd Intnl Conf. on Metaheuristics* (MIC'97).
10. Colorni, A., Dorigo, M. & Maniezzo, V. (1991). Distributed optimization by ant colonies. *Proc. First Europ. Conf. on Artificial Life* (Varela, F. & Bourgine, P., eds), pp. 134-142, MIT Press.
11. Colorni, A., Dorigo, M. & Maniezzo, V. (1992). An investigation of some properties of an ant algorithm. *Proc. of 1992 Parallel Problem Solving from Nature Conference* (Männer, R. & Manderick, B., eds), pp. 509-520, Elsevier Publishing.
12. Colorni, A., Dorigo, M., Maniezzo, V. & Trubian, M. (1993). Ant system for job-shop scheduling. *Belg. J. Oper. Res., Stat. and Comput. Sci.* **34**, 39-53.
13. Costa, D. & Hertz, A. (1997). Ants can colour graphs. *J. Op. Res. Soc.* **48**, 295-305.
14. Croes, G. A. (1958). A method for solving traveling salesman problems. *Oper. Res.* **6**, 791-812.
15. Deneubourg, J.-L. & Goss, S. (1989). Collective patterns and decision making. *Ethol. Ecol. & Evol.* **1**, 295-311.
16. Deneubourg, J.-L., Goss, S., Franks, N., Sendova-Franks, A., Detrain, C. and Chretien, L. (1991). The dynamics of collective sorting: Robot-like ant and ant-like robot. In: *Simulation of Adaptive Behavior: From Animals to Animats* (Meyer, J.A. and Wilson, S.W., eds.) pp. 356-365. Cambridge, MA: The MIT Press/Bradford Books.
17. Dorigo, M., Maniezzo, V. & Colorni, A. (1996). The Ant System: Optimization by a colony of cooperating agents. *IEEE Trans. Syst. Man Cybern. B* **26**, 1-13.
18. Dorigo, M. & Gambardella, L. M. (1997). Ant colony system: a cooperative learning approach to the traveling salesman problem. *IEEE Trans. Evol. Comp.* **1**, 53-66.
19. Farmer, J. D., Packard, N. H. & Perelson, A. S. (1986). The immune system, adaptation, and machine learning. *Physica D* **22**, 187-204.
20. Gambardella, L. M. & Dorigo, M. (1995). Ant-Q: a reinforcement learning approach to the traveling salesman problem. In: *Proc. ML-95, 12th Intnl. Conf. Machine Learning*, pp. 252-260 (Morgan Kaufmann, Palo Alto, CA).

21.   Gambardella, L. M., Taillard, E. D. & Dorigo, M. (1997). Ant colonies for the QAP. Technical Report IDSIA-4-97.

22.   Geake, E. (1994). How simple ants can sort out BT's complex nets. *New Scientist* **141**, 20-20.

23.   Guérin, S. (1997). Optimisation multi-agents en environnement dynamique: application au routage dans les réseaux de télécommunications. DEA Dissertation, University of Rennes I and Ecole Nationale Supérieure des Télécommunications de Bretagne.

24.   Huberman, B. A., Lukose, R. M. & Hogg, T. (1997). An economics approach to hard computational problems. *Science* **275**, 51-54.

25.   Kephart, J. O., Hogg, T. & Huberman, B. A. (1989). Dynamics of computational ecosystems. *Phys. Rev. A* **40**, 404-421.

26.   Kelly, F. P. (1995). The Clifford Paterson Lecture, 1995. Modelling communication networks, present and future. *Phil. Trans. R. Soc. London A* **354**, 437-463.

27.   Kirkpatrick, S., Gelatt, C. & Vecchi, M. (1983). Optimization by simulated annealing. *Science* **220**, 671-680.

28.   Koopmans, T. C. & Beckman, M. J. (1957). Assignment problems and the location of economic activities. *Econometrica* **25**, 53-76.

29.   Kuntz, P., Layzell, P. & Snyers, D. (1997). A colony of ant-like agents for partitioning in VLSI technology. In: *Proc. of 4th European Conference on Artificial Life* (Husbands, P. & Harvey, I., eds), pp. 417-424, MIT Press, Cambridge, MA.

30.   Lawler, E. L., Lenstra, J. K., Rinnooy-Kan, A. H. G. & Shmoys, D. B. (eds) (1985). *The travelling salesman problem*. Wiley.

31.   Lumer, E. & Faieta, B. (1994). Diversity and adaptation in populations of clustering ants. *Proc. of Third Intl. Conf. on Simulation of Adaptive Behavior*, pp. 499-508, MIT Press.

32.   Schoonderwoerd, R. (1996). Collective intelligence for network control. Engineer thesis, Delft University of Technology, The Netherlands.

33.   Schoonderwoerd, R., Holland, O., Bruten, J. & Rothkrantz, L. (1997). Ant-based load balancing in telecommunications networks. *Adapt. Behav.* **5**, 169-207.

34.   Steenstrup, M. (1995). *Routing in communications networks*. Englewood Cliffs, NJ: Prentice Hall.

35.   Stützle, T. & Hoos, H. (1997). The MAX-MIN ant system and local search for the traveling salesman problem. *Proc. IEEE Intnl. Conf. Evolutionary Computation* (ICEC'97), 309-314.

36.   Theraulaz G., Goss S., Gervet J. & Deneubourg J.-L. (1991). Task differentiation in *Polistes* wasp colonies: a model for self-organizing groups of robots. In: *From Animals to Animats*, Proc. of the 1st Intnl. Conf. on Simulation of Adaptive Behavior (Meyer, J.A. & Wilson, S.W., eds), 346-355, MIT Press.

37.   Theraulaz, G. & Bonabeau, E. (1995). Coordination in distributed building. *Science* **269**, 686-688.

38.   Wilson, E.O. (1971). *The Insect Societies*. Cambridge, MA: Harvard University Press.

39.   Wodrich, M. (1996). Ant colony optimization. BSc Thesis, Dept. of Electrical and Electronic Engineering, University of Cape Town, South-Africa.

Network Configuration Management in Heterogeneous ATM Environments

Bernard Pagurek[1], Yanrong Li[2], Andrzej Bieszczad[1], Gatot Susilo[3]

[1]Dept. of Systems and Computer Engineering
Carleton University,
1125 Colonel By Drive, Ottawa, Canada K1S5B6
email: pagurek@sce.carleton.ca
[2]Nortel, Ottawa
[3]Newbridge Networks, Ottawa

Abstract. . This paper covers the design and implementation of a generic model, based upon Mobile Agents, to perform PVC configuration management functions in multi-vendor ATM networks. The paper introduces an alternative or complement to existing ATM configuration management solutions. In it, an autonomous mobile agent travels from switch to switch to perform its configuration duties. A prototype implementation written in the Java programming language was developed for the validation of the proposed solution and was tested using a simulation testbed also written in Java.

1   Introduction

Mobile intelligent agents have introduced an interesting and exciting new paradigm for network management. This paper has chosen configuration management of Permanent Virtual Connections (PVCs) in Asynchronous Transfer Mode (ATM) networks to demonstrate the applicability and suitability of mobile agents for management purposes. Setting up and releasing PVCs is not a simple problem because each ATM vendor provides its own way to configure a PVC for its ATM switches and other devices. There is no standard or uniform way for the network operator to set up an end-to-end PVC in a heterogeneous ATM network.

The agents in existing network management systems tend to be monolithic and hard to modify, static, and often require substantial resources. Mobile agents do not statically reside on network devices, therefore can be created on demand and destroyed when no longer required. They tend to be substantially smaller than the agents in current network management systems are because they normally perform a single task. In their specialized tasks they can be capable of some degree of intelligence. In this case the mobile configuration agent can carry network state information with it, is capable of acquiring the information needed for its task from remote and/or local sources, and is capable of autonomously migrating to other nodes as needed. As a result mobile agents can substantially reduce the burden on the manager side because a large management task can often be divided into smaller manageable tasks which can be delegated to such mobile agents.

This paper proposes an innovative improvement to existing ATM configuration management solutions based upon the use of mobile agents and the Java programming language. Java was chosen for the implementation because of its platform independence and portability, because it supports code mobility, and because of its robustness. The paper consists of six sections. Section 2 summarizes briefly some relevant ATM information and some vendor specific information needed for a manager to establish a PVC. It does not attempt to describe ATM technology but it does indicate the variety of different PVC Management Information Bases (MIBs) in existence and then introduces the necessity of a common view of a PVC MIB in multi-vendor ATM networks. Section 3 reviews different approaches to PVC configuration management in heterogeneous ATM environments, including manager-to-manager based, CORBA based, and mobile code based. Section 4 describes the mobile agent infrastructure developed to facilitate the migration and control of mobile agents as well as inter-agent communication. Section 5 discusses communication with existing legacy management agents such as Simple Network Management Protocol (SNMP) agents already on a network node and how this applies to our problem. Aspects of PVC configuration using the mobile code architecture are examined here. Section 6 summarizes the paper's contribution and some future work.

2   Background ATM Knowledge

The understanding of this paper requires familiarity with some ATM terminology and some domain-specific knowledge. ATM[1] is a cell-switching concept used in networking. It uses a fixed length cell of 53 octets including a 5 octet header used for addressing and general administration of cells. There is an 8 bit Virtual Path Identifier (VPI) and a 16 bit Virtual Channel Identifier (VCI). The VPI and VCI are unique for cells belonging to the same virtual connection on a shared transmission medium. ATM operates in a connection-oriented mode. Before cells are transmitted from one user node to another, a logical/virtual connection setup phase must allow the network to perform a reservation of the necessary resources, for instance, bandwidth. There are two kinds of mechanisms to set up a connection, namely Permanent Virtual Circuits (PVC's) and Switched Virtual Circuits (SVCs). The former is pre-established manually via a network management station at each switch along the end-to-end path.

Telnet is often used to perform such tasks. The latter is set up on demand, based upon signaling procedures like Q.293. SVCs however, are not yet commonly available.

A network management system consists of four parts: a management station (or manager), an agent, a management information base (MIB), and network management protocols. Figure 1 illustrates the relationship among these parts in a traditional centralized management approach.

The management station serves as an interface for the administrator to the network management system. It translates administrator's commands into actual monitoring and control of the network elements. Various applications provided at the management station include end-to-end PVC provisioning.

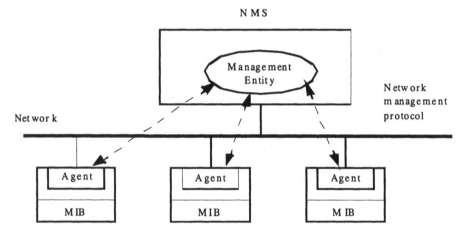

Fig. 1. Manager/Agent Model

Each node in the network that participates in network management contains agent software for performing management-related tasks, for instance, setting up a cross-connection at an ATM switch. Each agent collects data on resources and stores statistics locally. Agents also respond to commands from the manager. We note that these are classical communication and not AI agents.

The third component of the network framework is the MIB, a collection of objects, each representing a particular aspect of a managed device. For example, a manager can perform a monitoring function by retrieving the value of available bandwidth at a certain port of a switch, and changes the settings of the switch resource by modifying the value of available bandwidth.

Finally, the communication between the manager and agents is carried out using a network management protocol. The two standard network management protocols are the IETF's SNMP and ISO's common management information protocol (CMIP). In addition to defining a specific protocol, both SNMP and CMIP define a set of MIBs.

Due to the nature of design differences, each vendor currently develops its own device MIBs. For instance, Fore has its own SNMP MIB for its ATM switch products, as does Cisco. Attempts have been made to standardize such MIBs such as in the ATM Forum's ILMI 4.0 covering public and private User to Network Interfaces (UNIs). The IETF's ATM MIB (RFC 1695) is similar to the ATM forum MIB but seems more oriented towards PVC configuration and so is represented here. RFC 1695 defines a MIB used for managing ATM based interfaces, devices, networks and services. Its primary goal is to manage ATM PVCs [2]. Fig 2 describes the structure of this ATM MIB and shows that the MIB objects are divided into eight groups each containing variables (simple objects)

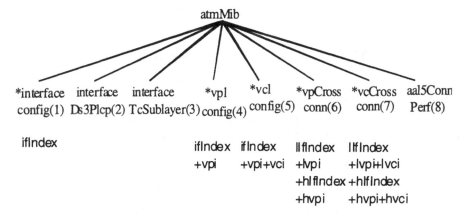

*denotes PVC configuration related group
and indexed tables of variables.

Fig. 2. IETF ATM MIB and Table Indices

For example the fourth group consists of the Virtual Path Link (VPL) parameters. A VPL is terminated in an ATM host or switch and can be created, modified, and deleted. The group has a table in which each row contains configuration and state information about a bi-directional VPL. Such a VPL entry is indexed by ifIndex and vplVpi the virtual path index for that VPL. This VPL group is similar to the Virtual Path Connection group in the ATM Forum's ILMI MIB.

Neither the ATM forum nor the IETF proposed MIBs have been adopted as standards, leading to a multitude of private MIBs. For a manager/operator to establish end-to-end PVCs, for each switch involved, he needs to know about

76

ports, QoS (quality of service required), bandwidth available, vpi and vci indicators, and how to establish cross-connects. It clearly would be advantageous if the operator could concentrate on a common view, which can be translated automatically by the appropriate interface into a form suitable for use with the individual vendor switches.

3 Different Approaches to PVC Configuration Management

Research activity in this area varies from integrating existing SNMP based configuration solutions to developing more innovative solutions such as using CORBA and Java. This section reviews several approaches for heterogeneous ATM networks.

3.1 Manager to Manager Based PVC Configuration Management

In the case of PVC provisioning, a domain is specific to a vendor and the managed objects like PVC cross-connections are in PVC related MIBs. A manager can access a managed object only through a related agent. Also, if a manager tries to access a managed object in a remote domain, it usually has to communicate and negotiate with a remote manager. This is shown in figure 3. For multi-domain operation, a multi-domain management protocol (MMP) has to be defined so those local managers can cooperate with each other to accomplish a global task like PVC provisioning.

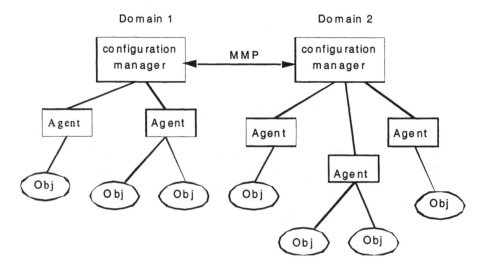

Fig. 3. Conceptual Model of Multi-domain

The advantages of this approach in a heterogeneous ATM network are:

- Use of existing manager software applications,
- Only requires implementing integration software between managers
- Rapid development Management

Because this approach uses existing software developed by the device vendors, there is no great need to develop new software, except the manager to manager communication software. It takes a relatively small amount of effort to integrate existing systems. The disadvantages of this approach are:

- Need different manager software from different vendors,
- Operator has to know all MIBs from different vendors,
- Integration becomes difficult when the number of vendors grows

Although this approach requires only the integration of existing management systems, it is required to have all the manager software applications available. Perhaps the biggest drawback is that integration among multiple vendors will inevitably be difficult.

It should be noted that there is another way to handle this kind of multi-manager system. Instead of using a peer manager-to-manager protocol, a top-level manger acting as central manager can be used to communicate with domain managers. These domain managers are mid-level managers and are the vendor specific products. In this model, a vendor specific PVC configuration management task is delegated to the mid-level managers by the top-level manager. The delegation is normally conducted using scripts. Unlike the manager/agent model, there is no standard protocol in SNMP for the communication between the top level managers and mid-level managers.

3.2   CORBA Based PVC Configuration Management

A second approach to PVC configuration management is use of the Common Object Request Broker Architecture (CORBA) [3]. The Object Management Group (OMG) initiated the Object Management Architecture (OMA) in recognition of the challenge of developing software applications and components that support and make efficient use of heterogeneous networked systems. CORBA is one of the key components of the OMA.   Two key components of CORBA are OMG's Interface Definition Language (IDL) and the IDL compiler. The IDL specification defines Object Oriented interfaces containing methods and attributes. It also maps IDL onto multiple programming languages, such as Java, C/C++, and Smalltalk etc.   The OMG IDL compiler generates procedure stubs on the client side and procedure skeletons or server stubs on the server side.

78

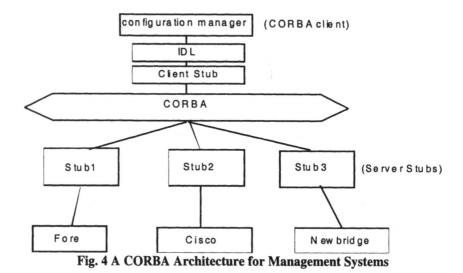

Fig. 4 A CORBA Architecture for Management Systems

Due to its heterogeneous capability, some research work has been done in applying CORBA in network management [4]. Figure 4 shows a possible management architecture in a multi-vendor ATM network environment. In the case of PVC configuration management, the IDL represents the PVC configuration common view, which is discussed in section 5.

The CORBA client or the end human user only needs to know the common view. The vendor specific switch MIBs are transparent to the client. Server stubs reside on the different devices themselves or possibly proxy machines. A client can invoke a method, say getAvailableBandwidth(), and each server stub translates the available bandwidth from the common view to its vendor specific MIB object. For instance, Stub1 translates the available bandwidth from the common view to the Fore MIB Object, " throughput".

Some advantages of this approach in a heterogeneous ATM network are as follows:

   - Deploying CORBA enabled Web browser,
   - Supporting legacy systems,
   - Coexistence with current manager platforms

Making use of CORBA follows the trend to develop Web-enabled software applications. With the wide use of the WWW, this feature is obviously attractive. Supporting legacy systems is perhaps even more important. That means that instead of getting rid of the existing systems, they can be evolved to CORBA, for example from SNMP. Another aspect of using CORBA is coexistence with current manager platforms, for instance, HP Openview.

Although the advantages of this approach are appealing, there are some disadvantages.

They are as follows:

- A "heavy weight" solution,
- Features depend upon services,
- Overkill for small applications like PVC configuration

That is, since CORBA is intended for large network systems, it primarily uses heavy weight protocols. This may cause network traffic problems. Although on top of CORBA's Object Request Broker (ORB), services like the naming service can be applied to quickly develop features such as "get/set a MIB object", these feature are normally expensive. The most important disadvantage of this approach is that for small application like PVC configuration, it is overkill to use CORBA because most of its features are not necessary.

3.3   Mobile Code Based PVC Configuration Management

Another way to approach PVC configuration management is to use Mobile Agents. This innovative approach is built upon a mobile code infrastructure discussed in the next section. As a practical application, mobile code based PVC provisioning tackles challenging issues like heterogeneity and interoperability. Figure 5 illustrates the general architecture of this approach. Discussion of how the code itself is handled and the internal details shown are, however, left for the next section. In this approach, assuming one of the simpler cases where best effort bandwidth allocation is sufficient, the PVC configuration manager sends a PVC agent to the switch at one end of the network connection. The agent performs the PVC configuration tasks like setting up a cross-connection between specified ports, and reserving bandwidth, vpi, and vci. Reading and setting specific variable values in the local MIB using the VMC as the interface does this. Then it migrates to the next switch in the route, carrying the available bandwidth and vpi/vci information needed by the next switch, where it performs similar duties. The agent traverses all the switches in the route until it finishes all configuration tasks, and can autonomously return to any node as needed. The agent can also return to the management station with the accumulated end-to-end information for display. More information on how this approach works is contained in Section 5.

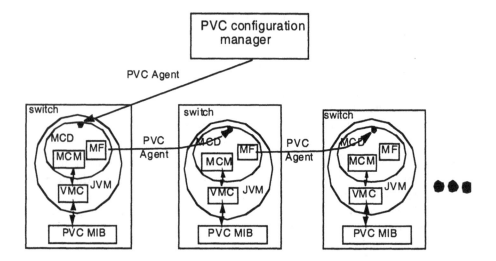

Fig. 5 PVC Configuration Management Based upon Mobile Code

3.4 Comparison of Methods

Compared with the previous two approaches, the mobile code based approach has outstanding advantages. First, there is no manager to manager software integration. The configuration manager has a common view of PVC configuration for all different switches. The operator does not have to be aware of the underlying switch systems from different vendors and is able to delegate responsibility to the mobile agent. Although vendor-specific MIB information is stored on each switch system, a switch interface, implemented as what we shall call a VMC (Virtual Managed Component) in the next section, is used to translate the common view to the vendor specific MIB information. The PVC agent automates the connectivity procedures without requiring a human operator's decision making. Secondly, the PVC agents are designed in such a way that they are specialized to perform only PVC related tasks. Therefore, compared to monolithic agents in conventional manager/agents model, they are much smaller.

The main justification for the agent approach is not efficiency but flexibility, extensibility, and ease of use. Current ATM switch agents are very difficult to change and extend when there is a need for them to adapt. Mobile code may easily be downloaded as needed, can readily be modified and so can be kept completely up-to-date. Mobile agent toolkits are becoming readily available while manager-to-manager protocols are not. It is our experience that Mobile agent toolkits are also fairly easy to apply.

While we have not yet been able to fully compare the agent approach from the efficiency point of view for this application, our experience leads us to believe that it will be competitive. It should be noted that while figures 3 and 4 seem to suggest that the configuration tasks can be done in parallel at all nodes simultaneously, this is not the case since information needs to be carried from one node to the next. Generally it is easy to get information from individual nodes but the mobile agent approach is very helpful in acquiring an overall or end-to-end picture which managers typically need. Current ATM agents are much too locally oriented to easily provide broader overviews.

4 Framework for Net Management with Mobile Code

Currently, there is much ongoing research using mobile agents by companies such as Crystaliz, Inc., General Magic, Inc., GMD FOKUS and International Business Machines Corporation [5]. In order to standardize mobile agents, the OMG (Object Management Group) proposed the Mobile Agent Facility Specification [6]. There are also a few applications using mobile agents to do things like providing management functionality where needed [7]. The Perpetuum Mobile Procura Project [8][9], which consists of infrastructure, simulator, network manager and application subprojects, is used to research the use of mobile code for managing networks.

4.1 Infrastructure for Mobile Code

A facility to transport portable code is a core requirement for solutions based upon mobile code. The infrastructure for mobile code requires a Java Virtual Machine (JVM) [10] running on each network component. A JVM is a platform dependent interpreter, which translates Java bytecode into machine code. At the time of writing, there are also JVM chips becoming available. Figure 6. shows the key components required to send mobile code from one network component to another.

Mobile agents can serve a variety functions and, depending on the application can have varying persistence[11]. For our purpose we make use of a deglet, an autonomous mobile agent delegated to perform a specific task and migrate within a limited region for a short period of time. The autonomy means that the agent has the intelligence to work independently of the manager and doesn't need to report back until the task is completed. It is even possible to imbed mini rule based expert systems in mobile agents. In fact such tools are available in Java implementations. This could be particularly useful when bandwidth or other resource limitations require an adjustment to the configuration plan.

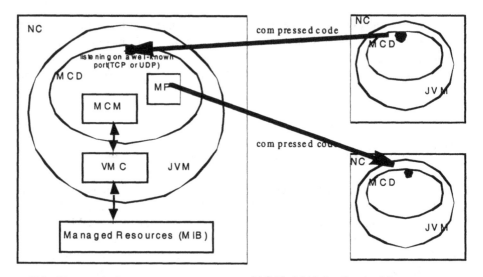

NC: Network Component   MCM: Mobile Code Manager
MCD: Mobile Code Daemon   VMC: Virtual Managed Component
MF: Migration Facility   JVM: Java Virtual Machine

Fig. 6 Infrastructure for Mobile Code

The MCD is the core component to realize code mobility in order to perform network management tasks. It provides a set of services that enable mobile code execution. These services include a runtime environment, a migration facility, an interface to access managed resources, and an agent-to-agent communication facility (not shown in figure 6). The MCD is a thread running inside the JVM and it listens at two communications ports to accept mobile code. One connection port is for TCP, while the other is for UDP. Once a piece of mobile code is accepted, it is instantiated within the same JVM as the MCD is running. The MCM looks after the actual management and storage of code keeping track of all handles of instantiated mobile code.

Next the code is security checked to ensure that it has not been tampered with and to authenticate the source. Then if it brings serialized state information with it, the state is restored on the new machine by invoking method onRestore and calling function onStart begins execution

The migration facility or MF, part of the MCD, is responsible for transporting a mobile agent to the next location. The application manager, the MCD, or the agent itself can decide the migration route. When an agent's migration is requested, the MCD calls method onMigrate(). This notifies the agent to be transported to the next location, giving it time to finish its current task. When the migration elsewhere is completed, the agent is destroyed and removed from the MCD. These details of the infrastructure operation are shown in the flow diagram of Fig. 7.

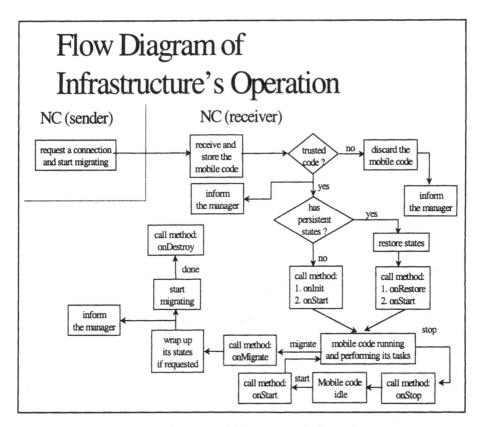

Fig. 7 Flow Diagram of Infrastructure's Operation

The ability to access managed resources of a network component is fundamental to network management. Accessing data in a heterogeneous system may be complex due to non-standard naming schema and procedures. Therefore, there must be a uniform way to be understood by the mobile code in performing its tasks without prior knowledge of the underlying system. The VMC interface makes it possible for mobile agents to access managed resources in a uniform way. In fact, the VMC concept is far more general and provides also for security methods, recovery procedures, translation, and provisioning procedures. The VMC acts as bridge between a mobile agent and managed resources. It is especially interesting because it can be updated at any time or downloaded on demand from a source URL, perhaps provided by the vendor. This feature provides for truly extensible agents.

Although not shown in the diagram, the MCD also includes a Communication Facilitator (CF) which provides communication between deglets. This is done through an external mediator, which keeps track of the location of active deglets.

5   Aspects of Mobile Agent Based PVC Provisioning

5.1   Interface to Legacy Systems

Many of the nodes will support a legacy static SNMP agent and MIB. For example the Fore switches support an agent with a proprietary SNMP MIB which can be accessed. To eliminate duplicating the effort of the SNMP agent in managing low level resources, it is possible to include full SNMP managerial type capabilities in a PVC provisioning deglet so that it can issue GETs and SETs and there are SNMP classes available in Java for this purpose [12]. This is a rather heavyweight approach, as it requires complete SNMP packet handling and BER encoding/decoding in the deglet. A simpler approach is to use a lightweight protocol such as the SNMP-DPI protocol (Distributed Processing Interface, RFC 1271) which allows for such communication in a simpler manner. This protocol was originally designed to provide SNMP agent extensibility through subagent technology. By adapting and implementing the DPI protocol in Java, in a symmetric way, full bi-directional SNMP type interaction between mobile agents and legacy DPI enabled SNMP agents was accomplished through the VMC. As a result, when SNMP support for PVC configuration is available in a node, it can be used avoiding the necessity of the VMC component having to duplicate the effort. If not, the VMC must provide the PVC configuration support. The way the Fore switch MIB is designed makes it impossible to use SNMP to configure a PVC. The available Newbridge switch does not support SNMP. In any event there is a variety of methods in use and it is necessary for the PVC deglet to come to grips with them.

5.2   Mapping PVC Common View to Vendor Specific MIB

The goal of setting up an end-to-end PVC based upon mobile code is to provide a uniform way for the ATM network operator to perform this operation. Thus he or she no longer needs to know each underlying switch MIB representation in a heterogeneous ATM network. In order to accomplish this, it is necessary to have a common representation, or common view of a PVC MIB. To this end the Forum's and IETF's MIBs as well as those of four different vendors were analyzed and a "common view" was abstracted. This was possible because, although different switches have their own ways to store data, they all perform more or less similar kinds of tasks in terms of PVC configuration. All the end user is aware of is the general switch PVC representation, or the "Common View". Because the actual PVC configuration needs to be done at each vendor specific switch, it is required to have a kind of mapping, or translation from the "Common View" to a vendor specific switch PVC MIB.

To materialize this mapping, a software component, called a Switch Interface, was developed based upon the mobile code infrastructure. It is a subclass of abstract class VirtualManagedComponent and acts as a bridge between a mobile PVC deglet, and a vendor specific switch MIB. It can be preinstalled in a switch MCD or downloaded on demand as needed. Therefore, when the PVC Configuration Manager wants to conduct some task, say setting up an end-to-end PVC, a PVC deglet with a "Common view" is launched by the application manager to a switch daemon at the start end of the network path. The deglet then gains access to the SwitchInterface. The SwitchInterface in turn obtains access to the underlying switch. After completing the task on one switch, the deglet migrates to the next switch.

5.3 Deglet Migration

Before the ATM cells can be transferred from one end of an ATM network to the other end, a virtual connection needs to be set up. In most cases, two end users make network connections through some network carrier. The end nodes represent the users (switch) and the intermediate nodes represent the carriers switches often called the carrier ATM cloud. From the network point of view, a carrier's ATM network can be considered as one switch, according to ATM Forum M4 [13]. Figure 5 with only three switches in total would depict this idea.

Prior to sending ATM traffic, current operation would be for the network operator at one end to configure the PVC at his or her switch. He or she then contacts the carrier's operator, either by phone or by e-mail. The carrier's operator configures its switch(es) based on the information given by the end switch operator. After that, he or she contacts the network operator at the other end. The operator at this end finally configures the switch at this end.

With the mobile code approach, only one operator is required. The human operator enters the end-to-end PVC configuration requirements, such as port connections for neighboring switches and bandwidth. The PVC configuration application, which serves as a network manager, sends a PVC deglet to the network to conduct the configuration task. The agent carries user requirements, accomplishes the configuration task at the switch 1 end first, then it migrates to the carrier's switch. It does the configuration task at the carrier's switch, then migrates to the switch 2 end. The agent finishes the task at the end switch 2, thus completing the end-to-end configuration task. It should be noted that the procedure is sequential rather than parallel. This is because in order to configure the carrier's switch or the end switch, a deglet has to finish the configuration on the previous switch first so that it can carry VPI/VCI values from one switch's outgoing port to next the switch's incoming port. Most of the foregoing discussion assumes the simplest case, that of UBR (Unspecified Bit Rate) traffic, which is the simplest from the point of view of quality of service. Only best effort bandwidth provisioning is required. In more stringent situations such as CBR (Constant Bit Rate) traffic, the QoS requirements are more complex and it is possible that the PVC configuration deglet might have to backtrack to accomplish its task. The autonomous migration capability for deglets in the mobile code environment allows the deglet to make intelligent migration decisions for such purposes. The configuration of virtual connections (VCs) using the IETF MIB is discussed in some detail in [14]. As can be seen the algorithms involved tend to be (in the authors' own words) fairly elaborate even without the complexity of proprietary definitions and notation.

6 Conclusions

6.1 Testing

Current ATM switches do not yet run Java Virtual Machines so actually setting PVCs in the real world would have required the use of proxy computers. To avoid this complexity, testing was conducted in a simulated ATM environment. To accomplish this, several JVMs, each with its own MCD were run on a few machines to simulate a small network. Setup was accomplished with a GUI and simulation management facility to establish the simulated elements. The deglets migrated from JVM to JVM much as in the real world. The switch environments and management databases were set up as VMCs. The configuration deglet performed as required, wandering through the network changing the appropriate proprietary MIB variables through the use of SwitchInterface to accomplish tasks such as setting up, tracing or releasing end-to-end PVCs. The difference, of course, was that the switches were not represented in a dynamic way, with cells flowing through them, but as a MIB to be read and written to by the visiting deglet.

6.2 Summary

This paper has presented a new approach to PVC provisioning in a uniform way through the use of mobile agents called deglets. In order to overcome requiring knowledge of the PVC configuration MIB for each individual switch in an ATM network for network operators, the approach introduces a "Common view". With this PVC configuration "Common View", a network operator does not have to be aware of PVC configuration MIBs for different vendor switches. A software component acts as an interface to map the "Common View" to each vendor specific PVC configuration data.

A prototype implementation of the software application was created. It consists of twenty-two classes and 6000 lines of Java code. With its graphical user interface, which is what a network operator would see, a user can choose among the different PVC configuration functions. The underlying software launches a specific mobile deglet depending upon the user's choice of PVC configuration tasks.

The implementation of management features using the mobile agent approach reduces the complexity of the implementation. The creation of a single purpose agent reduces the interactions among network management features, which can only reduce implementation time. This single purpose agent, because it is small, is also likely to be smaller than the data transferred among nodes as compared to the monolithic agent approach.

6.3 Future Work

The current program is based upon simulating a multi-vendor switch ATM environment. The software needs to be refined so as to accommodate real ATM switches. If this requires the use of proxies, it will be done so that we can also implement the CORBA approach. It will be useful then to compare the various approaches from the point of view of:
- Ease of use;
- Ease of implementation;
- Efficiency and network traffic generated.

Broader classes of QoS need to be explored in terms of the reasoning and migration decision making required. To facilitate reasoning and to make algorithm implementation easier we are introducing a forward reasoning rule-based component into our agents. Finally, the PVC tracing and monitoring functions are being augmented with an alarm correlation facility using a codebook approach.

Acknowledgements

This work was supported by the Telecommunications Research Institute of Ontario and the National Science and Engineering Research Council of Canada.

7 References

1. Martin DePrycker, Asynchronuous Transfer Mode: Solution For Broadband ISDN, EllisHorwood, 1995
2. Ahmed M., and Tesink K., " Definitions of Managed Objects for ATM Management Version 8.0 using SMIv2" RFC1695, IETF, August 1994.
3. The Common Object Request Broker: Architecture and Specification Rev. 2.0 , Object Management Group, Framingham Mass., July 1995

4. Leppinen M., et al, "Java- and CORBA-based Network Management" IEEE Computer, June 1997, pp83-87.
5. Kiniry J., and Zimmerman D., " A Hands on Look at Java Mobile Agents" IEEE Internet Computing, July-August 1997, pp21-30.
6. Mobile Agent Facility Specification, OMG TC Document cf/xx-x-xx, June 1997.
7. Busse I., and Covaci S., " Customer Facing Components for Network Management Systems, Integrated Network Management V, Proceedings Fifth ISINM, San Diego, May 1997, pp31-43.
8. Carleton University Perpetuum Mobile Procura Project, http://www.sce.carleton.ca/netmanage/perpetuum.shtml, 1997
9. Susilo G., Bieszczad A., and Pagurek B, " Infrastructure for Advanced Network Management Based on Mobile Code", Proceedings of the IEEE/IFIP Network Operations and Management Symposium, NOMS '98, New Orleans February 1998
10. The Java Virtual Machine Specification, Tech. Report, Sun Microsystems, 1996
11. Bieszczad A., and Pagurek B., "Network Management Application-Oriented Taxonomy of Mobile Code" Proceedings of the IEEE/IFIP Network Operations and Management Symposium, NOMS '98, New Orleans, Feb. 1998.
12. Advent Network Management Inc. "Advent Network Java SNMP Package", http://www.adventnet.com/snmp_api.html, 1996.
13. ATM Forum Technical Committee, M4 Interface Requirements and Logical MIB, ATM Forum, October 1994
14. Tesink K, and Brunner T., "(Re)Configuration of ATM Virtual Connections with SNMP" The Simple Times, Vol.3, Number 2, August, 1994. st-subscriptions@simple-times.org

Agent-Based Schemes for Plug-and-Play Network Components

Andrzej Bieszczad, Syed Kamran Raza, Bernard Pagurek, Tony White

Systems and Computer Engineering, Carleton University,1125 Colonel By Drive, Ottawa, Ontario, Canada K1S 5B6

{andrzej,skraza,bernie,tony}@sce.carleton.ca

Abstract. In this paper, we present several approaches to making the process of configuring network devices easier than is currently the case. Configuring a device requires that a number of attributes in the network and on the device be set as well as certain software components be installed. For example, a printer requires that its drivers are present on workstations that will be using it. Currently, the manager of the network has to perform all required tasks manually. We describe four approaches that start with a procedure similar to the way plug-and-play components are handled by operating systems; e.g. Microsoft Windows. We extend the plug-and-play idea to network components. Next, we describe how this basic routine can be extended to a simple distributed model. A natural extension of the distributed model is the use of CORBA to handle a distributed environment. Finally, we propose a model of plug-and-play-ness based upon mobile agents and describe its implementation. We argue that the last approach has a number of advantages over the other models.

1 Introduction

Telecommunication networks comprise large numbers of heterogeneous devices. The challenge of managing components starts when a device is purchased and needs to be added to the network. Establishing a hardware connection is the first step, which requires prior research on available interfaces including physical connections and low level communication. It would be hard to imagine that this process can be automated. The remaining installation requires configuration of the device, the network and activation of appropriate drivers. This provisioning process, which is traditionally performed by network managers, is time consuming and frustrating, because very often the devices require software elements that are not immediately available. On a smaller scale, let us imagine a computer user trying to install a CDROM drive on his computer. If the drive comes with the right driver, then the installation is relatively straightforward. A casual user will still need to obtain advice from experts, but in the end, he or she will have his CDROM drive working. The problem starts when the drive does not come with a right driver. It might be an outdated driver or a driver for another operating system. To address these kinds of problems, vendors of operating systems and hardware components developed the idea of plug-and-play-ness. For example, the Microsoft Windows operating system includes a collection of drivers that can be installed automatically for a device that has been recognized by the system. If a device is recognized, then the installation process is almost transparent. We say 'almost' because the system usually has to be restarted, and this requires attention of the user. The trouble starts when there is no match for the device in the database.

At this moment, there are no similar solutions available in network management systems. There are several reasons for that. The interaction between a specific device and the network is relatively complex. A registry of all devices and their interaction schemes would be very large. Additionally, provisioning a network device is usually more complex than installing a driver, so a plain database might not suffice. Another problem is that there are no operating systems to handle automatic installation processes in networks. In spite of all of these problems, applying ideas rooted in the PC world might still ease the provisioning task considerably.

In the remaining parts of sections paper, we analyze several approaches that can be used to implement plug-and-play network components. First, we adapt the simple plug-and-play installation process based on a local database of drivers to the networked environment. Next, we extend the basic model by proposing that the driver repositories be remote and distributed. In the third model, we apply concepts from open distributed computing environments to link the device being installed with the local environment and with remote repositories. Finally, we introduce a solution based on mobile agents. We list the advantages and disadvantages of each of the schemes. We argue that the solution based on mobile code is superior to all the others presented. We base these opinions on our experiences in applying mobile agents in the Network Management domain.

The Perpetuum Mobile Procura group at Carleton University focuses on addressing many of the issues in traditional network management systems with techniques based on mobile code [0]. Our ultimate goal is a plug-and-play network (PnPnet) [0]. Plug-and-play components play an important role in our vision of future self-configuring, self-diagnosing and self-repairing networks. Before we move on to the schemes for plug-and-play components, we briefly introduce the mobility framework that we are using in our research.

Using only one term agent to refer to all instances of mobile code is frequently confusing. This problem is exacerbated, if we take into account that the same term is heavily used in the context of network management. That is why we have proposed a taxonomy of mobile code [0], which we use throughout this paper.

2 Plug-and-play strategies

2.1 A scheme with local repositories

The first scheme for providing a plug-and-play capability for network devices is based on solutions designed by vendors of operating systems. In this scheme (**Fig. 1**), the data and procedures that are required by a device to operate in a network are stored in repositories maintained by network managers. These repositories have to include the elements that the network needs to accommodate the device. A repository has to be prepared prior to the arrival of a new device. A special process running as part of the operating system of the management workstation handles the installation. We can consider this process to be a network plug-and-play agent. A management workstation could be any device that can communicate with new devices. The network plug-and-play agent has to implement a communication protocol that needs to be handled on both sides, the device and the management workstation providing plug-and-play services. The handler of this communication on the device can be viewed as component plug-and-play agent. Furthermore, the network plug-and-play agent needs access to a repository of provisioning data. The network plug-and-play agent is programmed to repeat the same scenario for every new device. First, after a new device has been successfully connected to the network[1], the network plug-and-play agent has to detect its presence, so the initialization process can be started. For example, polling of connected devices must be performed analogously to the way in which an operating system scans the components connected to the bus during the boot process. In networks, a discovery algorithm must be run at some other time, because networks are shut down infrequently. The next step is to activate an identification routine, in order to determine the type of the component. Then, the repository of configuration routines is searched for the provisioning modules that correspond to the detected device. The component plug-and-play agent retrieves the necessary items from the repository and installs them accordingly. It may negotiate with the network plug-and-play agent to determine all necessary details of the installation.

---

[1] We handle this initial connection with greater care in the section devoted to the scheme based on mobile agents.

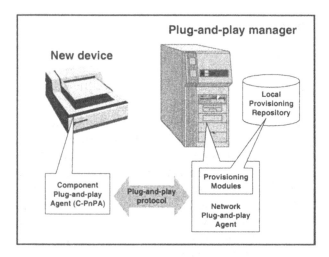

Fig. 1. Plug and Play with Local Repository

This scheme has a number of advantages and disadvantages when compared to manual provisioning. These are:

2.1.1 Advantages
- Plug-and-play capabilities ease the task of network managers, who have to deal with dynamic network topologies. The network plug-and-play agent provides a scheme for auto-provisioning new components. A network operator does not need to know all procedures and data that are required to install dozens of network components.
- The scheme is relatively efficient, because it involves local repositories, and local communication.

2.1.2 Disadvantages
- Every network has to maintain repositories of all potential devices.
- The database has to be available from multiple servers to provide a degree of fault tolerance. This also constitutes a synchronization problem.
- A far more serious problem is the need to keep the database up to date; i.e. to synchronize it with the evolving hardware and software.
- Each device has to implement plug-and-play negotiation protocols. They might be different for different networks. The only solution is to establish a standard protocol, because neither a device can include protocols for every network, nor every network can implement protocols for every device possible.
- Even with a standard plug-and-play protocol, the plug-and-play agents remain platform-dependant. This constitutes a major maintenance problem, because each device needs its own version of an agent. The same applies to the networks.

- There is a need for a discovery process running constantly on a dedicated resource in the network. This process consumes resources even if the network topology does not change.
- After the installation process has concluded, the only way to repeat it is by operator intervention. There are no provisions for automatic updates to the current settings, even if the database is modified by some means.

2.2   A scheme with remote repositories

The next scheme that we analyze is an extension of the previous model. The idea again comes from the PC world, although it is not yet wide spread. Vendors realized that maintaining repositories of data locally constitutes a problem for the user. After the original purchase of their product, the database quickly becomes obsolete. In this extension, certain details are brought from remote databases, which are maintained by the vendor of the operating system.

In the networked world, we do not have an operating system, so the problem is more complex. For a vendor of an operating system it is easy to synchronize its programs and repositories. In a network, we have to deal with a multitude of hardware and system vendors. As in the first scheme, we need a network plug-and-play agent, which runs on one of the network computers and implements the plug-and-play protocol (**Fig. 2**). In contrast to the previous model, however, the provisioning modules are obtained from remote databases. Each vendor needs to maintain a repository of provisioning modules. Before the network operating company purchases a device, the address of the repository (e.g. URL) must be stored on the device. During the exchange of messages using the plug-and-play protocol, the network agent retrieves the address. It is designed to contact the vendor's repository and obtain all required provisioning modules. They are brought to the network and distributed according to the installation procedure.

Installation procedures may vary between various components. Each of them is a combination of network- and device-specific routines. Accordingly, the network portion can come from the network agent, and the device part can be brought from the vendor's repository.

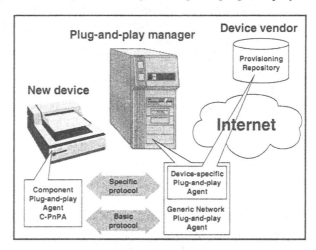

Fig. 2. Plug and Play with Remote Repositories

This device-specific plug-and-play agent would be activated after the generic part of the protocol recognizes the type of the device. Both agents cooperate and negotiate to complete the installation. After the installation, the device agent can be terminated and removed from the system.

The scheme involving downloadable code is increasingly popular amongst software vendors. Several products are distributed only as anchors for downloading remaining software components from the vendor's repositories. To update software, the user either needs to revisit vendor's Web sites that contain checking agents or a local agent is run every time the software is run. The scheme works in the environment with direct connection to the Internet. For example, Microsoft uses a similar routine for installing Internet Explorer version 4. Others, like Netscape and RealPlayer, implement the update schemes. All of these schemes involve native, platform-specific code.

2.2.1 Advantages
- Relegating the maintenance task to the vendor of a device solves the update problem. Vendors can keep the newest provisioning modules in their repositories.
- With the installation routine divided into two parts, it is not necessary to store installation algorithms for all possible devices in the network. A small generic agent is easier to maintain.

2.2.2 Disadvantages
- The downloadable device plug-and-play agent has to be able to run in any network execution environment, so a platform-specific version is required. Each vendor needs to provide native versions for all networks in which the device may potentially be installed.
- The security of the network is endangered by downloadable code. Although an authorization procedure might be in place, the network is not able to control the execution of native code after it passed the security checks.
- There is no mechanism for discovering new components. The managing process has to perform polling.
- There are no provisions for handling network dynamics and reconfiguration. For example, a newly added workstation may need drivers for a printer that was installed before.

2.3 A scheme based on CORBA

The Common Object Request Broker Architecture (CORBA) is a set of specifications for open distributed computing environments [0, 0]. Since the specifications were agreed, many CORBA implementations have appeared and been applied in various domains; e.g. Network Management [0].

CORBA allows for cooperation of remote objects. Applications can be easily distributed, because CORBA provides location transparency in method invocations. That is, a method implemented by a remote object can be invoked in the same way as a local one. The objects that constitute an application are connected to a data bus, or Object Request Broker (ORB). CORBA hides platform dependencies of participating objects. The Internet Inter-ORB Protocol (IIOP) is used[2] to invoke methods, send parameters and receive results.

Plug-and-play components can be implemented using CORBA (**Fig.** 3). The installation process is implemented as a distributed application

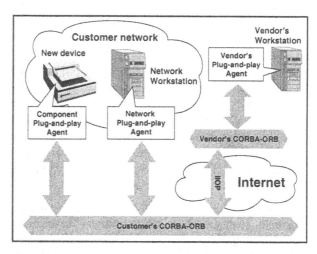

Fig. 3. Plug and Play using CORBA

with several objects residing on the new device, and on the network and vendor's resources. We will refer to the objects residing on the device as component plug-and-play agent. The objects residing on the vendor's premises constitute the vendor's plug-and-play agent. The network plug-and-play agent is a set of objects and services residing in the network. All agents are connected to CORBA-compliant ORBs. IIOP is used to link the ORBs. Agent objects communicate standard interfaces defined in Interface Definition Language (IDL) [0]. All applications that want to use the interface have to include stubs that are obtained by compiling IDL specifications[3].

---

[2] IIOP does not have to be used between objects residing on the same ORB. However, it is required to interact with ORBs of other parties. Every CORBA-compliant ORB must implement IIOP. In the context of this application, IIOP delivers transparency between calling methods local to the network, and those residing on vendor's premises. Usually network objects are connected to a different ORB than vendor's objects.

[3] A more complex technique based on Dynamic Invocation Interface (DII) [0] can also be used, but we will not analyze it in this paper. In this technique, the details of object interfaces are obtained from dynamic Interface Repository (IR). For example, Component plug-and-play agent could supply Network plug-and-play agent with a reference to an object on Vendor plug-and-play agent. Then, the reference could be used to obtain the details of the object's dynamic interface from the IR. Objects implementing dynamic interfaces have to register these interfaces with CORBA, which stores them in the IR.

After the new device has been connected to the network, a component plug-and-play agent is started during the bootstrap process. It uses a CORBA Naming Service to locate the network plug-and-play agent. The network plug-and-play agent must implement a standard PnP interface, so every component can use methods from this interface to exchange provisioning information with the network. The component plug-and-play agent has also a standard interface that the network plug-and-play agent can use to configure the component. The CORBA Security Services are used to ensure that the integrity of the network remains uncompromised. A component plug-and-play agent can also use methods on a vendor plug-and-play agent; e.g. to obtain provisioning modules that are appropriate for the environment in which the component is being installed. The interface between a component plug-and-play agent and a vendor plug-and-play agent does not have to be standardized, because these agents belong to the same vendor. However, if a network plug-and-play agent is requires access to functionality provided by vendor plug-and-play agent, then it has to be done through a standardized interface[4].

Let us summarize. In this scheme, the installation process is realized through a partially standardized series of reciprocal invocations of methods on objects of the participating agents. The rules of communication between the component plug-and-play agent and the network plug-and-play agent together with the related interfaces can be considered a standard plug-and-play protocol.

2.3.1 Advantages
- The most important improvement is platform independence. In this solution, one version of a network plug-and-play agent serves all devices, and one version of a vendor plug-and-play agent serves all networks.
- It is easy to upgrade vendor plug-and-play agents, as long as the interfaces are left intact.
- The new component can initiate installation, so there is no need for a discovery process.
- A number of network plug-and-play agents can be distributed in the network, so no single point of failure is present.
- New or updated provisioning modules can be uploaded by the vendor plug-and-play agent when they become available, but it requires that permission be granted by the network plug-and-play agent.

2.3.2 Disadvantages
- CORBA is required on each device. This is very expensive and might not be acceptable for small devices.
- New or modified vendor plug-and-play agent interfaces can be provided only through DII (See footnote 3), which is complex.
- It is not possible to upgrade a component plug-and-play agent.

---

[4] DII is an option. See footnote 3.

- Polling is required for new devices that depend on the services provided by a device. It can be handled by the component plug-and-play agent, or delegated to the network plug-and-play agent.
- Communication between remote CORBA objects depends on the ability to establish and maintain a healthy connection.

2.4 A scheme based on mobile code

The plug-and-play scheme shown in **Fig. 4**, which is based on mobile code assumes that an infrastructure for mobile agents [0] is in place. For example, the Java-based Mobile Code Toolkit, MCT is a framework implemented by Carleton University's Perpetuum Mobile Procura group [0]. The Open Management Group's Mobile Agent Framework (MAF) is an effort to establish a standard that would ensure interoperability of mobility toolkits from various vendors [0]. Furthermore, migration patterns should be established between the participating mobile agent daemons through their agent migration facilities. The network must be configured for plug-and-play-ness by offering not necessarily mobile plug-and-play agents in at least one location in the network. The network PnP agent must be accessible through a standard logical identifier.

While expecting a software implementation of Java [0] Virtual Machine (JVM) to run on every possible device might be too much, it is conceivable that small devices will utilize a JVM on a chip (e.g. PicoJava).

The installation process of a new device starts at a lower level of the communication hierarchy. A protocol such as the Dynamic Host Configuration Protocol (DHCP) [0] is used to assign an IP address and other required details; e.g. address of a DNS server, and subnetwork mask. After bootstrapping, the device is able to communicate with the network

First, the migration facility of the mobile code daemon residing on the new component is incorporated into the mobile agent infrastructure spanning the network. Establishing migration patterns requires not only that an IP address be assigned to the new component, but also that the migration facility on the device receives a migration target in the

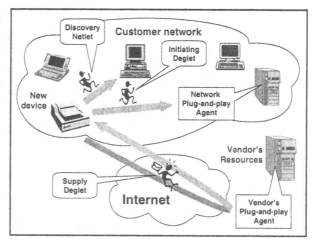

Fig. 4. Plug and Play using Mobile Agents

network. At least one IP address that can be used as a default migration target for the new device is obtained using a standard logical identifier of the network PnP agent. That is achieved using the naming facility. The naming facility is capable of forwarding the agent to the intended destination using a logical identifier as the target, as explained earlier. The bootstrapping agent sends an initiating delegation [0] agent (deglet [0]) to the network plug-and-play agent using its logical identifier. This deglet must be allowed to reconfigure the migration facility of one of the mobile code daemons in the network to accommodate the new device. For example, in a way analogous to inserting a new element in a linked list, the migration facility of the new device is set to point to the target of the selected network mobile code daemon. The migration facility on this selected mobile code daemon is then re-arranged to point to the new device.

Thanks to device behavior extensibility using extension agents (extlets [0]), the initial configuration of the mobile code daemon of the new component can be minimal. After the device is connected to the network, the mobile code daemon serves as an anchor for downloading a number of extlets from the vendor's repositories. The basic code transport facility of the mobile code daemon can be used to download all required extension agents from the vendor's repository. In an even more futuristic vision, extlets might also be used to engineer the actual functionality of the device. For example, only the extlets comprising the functional modules that were paid for would be downloaded.

The second delegation agent[5], a discovery deglet, is also dispatched to the network PnP agent to discover configuration requirements and negotiate the configuration terms applying to the whole network. This agent can be sent using the logical identifier as before. If the former delegation agent obtained the IP address of the network plug-and-play agent, then the discovery deglet can be sent directly. At the same time, one or more permanent discovery agents (netlets [0]) are injected into the network. It is possible, because the new device has been integrated to the migration web of the network's agent infrastructure. These netlets will ultimately visit all network components. The task of the discovery netlets is to obtain configuration requirements of individual components of the network. They stay in the network, so the device is informed about any future changes to the network. Should any such change require a modification to the configuration of the device or the network that affects the device, an appropriate update procedure will be performed. For example, a discovery netlet can detect all workstations that need drivers for a newly installed printer. If a new workstation is added in the future, it will also be accommodated.

In the next step of the configuration process, a supply delegation agent is sent to the vendor of the device (the IP address has been supplied with the device). Although a proprietary interface could be used between the supply delegation agent and the vendor's PnP agent, we propose that a standardized interface be

---

[5] It could be the same agent, but we want to keep our agents small.

used instead. This solution allows for third party provisioning modules. For example, a printer vendor might redirect requests for drivers to a third party.

The task of the supply delegation agent is to bring all required provisioning modules back to the network. They can be distributed to their destinations by another delegation agent. For instance, in the example with a printer, the supply delegation agent would bring drivers for all types of workstations discovered in the network. These drivers would then be distributed and installed.

The supply deglet may leave a trace of the printer in the registry maintained by the vendor's PnP agent and ask that the printer be notified of changes to the registry. If new provisioning modules are available, they can be sent to the device. Through the permanent assignment of the discovery netlet, the device knows what and where should be updated. If a new network component is discovered that requires provisioning modules that have not been brought yet, then another supply deglet travels to the vendor plug-and-play agent.

With the mobility framework in place, there are opportunities for other uses of mobile agents after the device is installed. For example, device-specific management applets can be stored on the device or brought from the vendor's repository [0]. A management applet can handle the device on the management workstation, which provides a docking service. The applet can provide coherent graphical displays and facilities to interact with the device.

Another opportunity unique to the use of mobile code for plug-and-play-ness is a possibility of using extlets and servlets to install certain services that the device requires in the network. This potentially very powerful capability would only be possible if the network grants appropriate permission.

We claim that the plug-and-play scheme based on mobile code is superior to the others presented in this paper. There are several advantages and disadvantages with this scheme. These are:

2.4.1   Advantages

- The scheme is platform independent.
- A new device initiates the installation process, so there is no need for polling.
- An ongoing automatic discovery performed by discovery netlets allows for automatic dynamic reconfiguration.
- Updates and modifications to the provisioning modules are brought from the vendor and installed in the network automatically.
- A dynamic network of distributed plug-and-play services is possible. For example, third party enterprises might compete for the network configuration business.

- Any agent can be upgraded easily. Firstly, its old version can be intercepted and killed. Secondly, an alive agent can be modified through the class-loading feature of Java.
- The scheme enables extensibility of device behavior through extlets.
- Extlets or servlets can also be used to install handlers of certain aspects of the device on the network.
- Migration facilities may use UDP to transfer agents in unreliable networks.

2.4.2 Disadvantages

- A framework for mobile agents must be installed in the network. In our case, it implies that each network component must be Java-enabled or has a Java-enabled proxy.
- There is a potential security risk. This is a problem inherent to any application based on an open system using mobile code. We recognize the importance of this issue and we are working towards overcoming the most serious problems.
- Discovery netlets are always alive, so they consume processing resources and lower the throughput of the network. However, the number and size of netlets is almost negligible in relation to the volumes of information transferred over an average network. We are working on density control mechanisms that will regulate the number of netlets present in the network.
- Transferring agents over TCP requires a healthy network.

3   Conclusions and future work

In this paper, we have presented several schemes for plug-and-play network components. Plug-and-play components are important in our vision of future trouble-free networks. Through a number of lists itemizing advantages and disadvantages of each of the methods, we have supported our claims that the scheme based on the use of mobile code is superior to all others presented in this paper.

Currently, we are actively implementing a scheme for configuring network printers based upon mobile code. We believe that in the course of the implementation process that we will define an ontology that can be generalized to accommodate any networked device.

Acknowledgments

This work is supported by grants from Communication Information and Technology Ontario (CITO) and Natural Sciences and Engineering Research Council of Canada (NSERC).

4 References

Bieszczad, A. and Pagurek, B., (1998), *Network Management Application-Oriented Taxonomy of Mobile Code*, to be presented at IEEE/IFIP Network Operations and Management Symposium NOMS'98, February 15-20, 1998, New Orleans, Louisiana.

Bieszczad, A. and Pagurek, B. (1997), *Towards plug-and-play networks with mobile code*, in Proceedings of the International Conference for Computer Communications ICCC'97, November 19-21, 1997, Cannes, France.

Dynamic Host Configuration Protocol, Draft Standard RFC 2131, Internet Engineering Task Force (IETF), March 1997. ftp://ftp.isi.edu/in-notes/rfc2131.txt

Gosling, J. and Arnold, K., Joy, B., Steele, G., Lindholm, T., Walrath, K., Campione, M., Yellin, F. et al, *The Java Series*, Addison-Wesley, 1996.

Hurst, L., Cunningham, P. and Sommers, F., *Mobile agents - smart messages*. In Proceedings of the 1ˢᵗ International Workshop on Mobile Agents, Berlin, Germany, April 1997.

Leppinen, M., et al., *Java- and CORBA-based Network Management*, IEEE Computer, June 1997, pp83-87.

Mobile Code Bibliography, http://www.cnri.reston.va.us/home/koe/bib/

Object Management Group, *Mobile Agent Facility Specification*, OMG TC Document cf/xx-x-xx, June 1997.

Object Management Group, *The Common Object Request Broker: Architecture and Specification Rev. 2.0*, Framingham Mass., July 1995.

Schramm, C., Bieszczad, A. and Pagurek, B. (1998), *Application-Oriented Network Modeling with Mobile Agents*, to be presented at IEEE/IFIP Network Operations and Management Symposium NOMS'98, New Orleans, Louisiana, February 1998.

Susilo, G., Bieszczad, A. and Pagurek, B. (1998), *Infrastructure for Advanced Network Management based on Mobile Code*, to be presented at IEEE/IFIP Network Operations and Management Symposium NOMS'98, New Orleans, Louisiana, February 1998. Also available as Technical Report SCE-97-10, Systems and Computer Engineering, Carleton University, Canada, May 1997.

Vinoski S., *CORBA overview: CORBA: Integrating Diverse Applications Within Distributed Heterogeneous Environments*, IEEE Communications Magazine, Vol. 14, No. 2, Feburary, 1997.

Yemini, Y., Goldszmidt, G. and Yemini, S., *Network Management by Delegation*, The Second International Symposium on Integrated Network Management, Washington, DC, April 1991.

Dynamic Resource Allocation by Market-Based Routing in Telecommunications Networks

M.A. Gibney and N.R. Jennings

Department of Electronic Engineering, Queen Mary and Westfield College,
University of London, London E1 4NS, UK.
{M.A.Gibney, N.R.Jennings}@qmw.ac.uk

Abstract. We present an approach to resource allocation in telecommunications networks based on the interaction of self-interested agents which have limited information about their environment. A system architecture is described which allows agents representing various network resources, potentially owned by different real-world enterprises, to coordinate their resource allocation decisions without assuming a priori cooperation. It is argued that such an architecture has the potential to provide a distributed, robust and efficient means of traffic management for telecommunications networks. Some preliminary work on the design of the trading behaviour of the agents in the economy is presented, including the results of experiments which investigate the relative performance of market-based agents compared with traffic management based on static routing.

1   Introduction

In a telecommunications network, a call between two parties may be connected via one of a number of paths. The process of deciding which of these paths to use is called *routing*. Choosing an efficient path is important because the network's capacity for handling traffic is finite, and when it is saturated, calls have to be turned away. This constitutes a loss of income to the network provider. However, finding the optimal path is problematic because the network state continually evolves. By the time the information needed to compute the optimal path between any two nodes is made available at the node where that decision needs to be taken, the network state will probably have changed, rendering that decision obsolete. Furthermore, efficient routing decisions, those which maintain a balance in utilization of the network resources, require information about the utilization of all network resources to be made simultaneously available to the process making that decision.

When a call is requested in a circuit switched network, a call set-up or *connection admission process* tests paths across the network determined by its routing algorithm for congestion and assigns the call to the first path which is uncongested. Routing algorithms are used to establish the appropriate routing paths or the equivalent routing table entries in each node along a path. Most algorithms are based on assigning a cost measure to each link in the network and determining the linear sum of paths across the network [1]. Based on these costs, the network tries to allocate traffic to the cheapest paths across the network. Where the cost function is based on link congestion (real or predicted) the cheapest path may change over time to maintain the network level efficiency of the call admission policy. However, such mechanisms are limited by their lack of information about the wider network state. Put simply, a connection admission process cannot determine the most efficient path from the network point of view merely by checking a small number of paths for congestion. The usage of the links in a path may be equal in two otherwise different network states and the most efficient routing decision from the network point of view would be different in each case. To rectify this situation, we propose a network management framework in which the effects of routing decisions in one part of the network are felt across the entire network, and in which node level decision-making takes place in the presence of some (limited) information about the network level state of congestion. Our approach is to model the telecommunications network resources as trading entities and goods in a computational economy. Although our investigation into this approach is still at the proof of concept stage, we show in this paper that a market-based mechanism is able to successfully route calls through a network over time, although not yet as efficiently as a conventional static routing mechanism. In this paper we discuss the relative performance of the two mechanisms under a variety of network loading conditions and the possible reasons for the under performance of the market-based approach. .

One of the main strengths of a market-based approach is that it provides a framework in which the various network components can be owned by different, competitive, real-world enterprises. Thus resource allocation takes place against the assumption of competition, rather than cooperation between the components. - this theme is discussed in section 2. In section 3 we describe the choices that were made in specifying the system architecture and the institutional structure of the market and how these follow from the system level properties we want to achieve. Section 4 describes the design of the individual agents. Section 5 discusses experiments devised to measure the performance of the market-based routing mechanism in a small simulated network under conditions of low, medium and high loading. In section 6 we compare and contrast our approach to that of others who have used market-based mechanisms for network control problems. Finally section 7 describes the open issues and future work.

2   Background and Motivation

A number of agent-based approaches to resource allocation and load balancing in telecommunications networks, for example [2],[3], have worked from the premise that the ideal resource allocation mechanism is one which considers the network as a single resource. In such cases, the system manages the network so as to optimise utilization of that resource. These approaches assume that since global load balancing is a common good, agents should be modelled within a co-operative framework. However, it is not clear that the model of single ownership implied by this co-operative framework will continue to dominate telecommunications deployment in the future. As an economic concern, a network derives its value from connectivity. Thus, large networks grow by acquisition and small networks are swallowed up. However expansion can also happen through co-operative agreement between rival networks to provide inter-connectivity. In the latter case, larger multi-owner networks are created. Another increasingly common aspect of modern telecommunications deployment is the practice of enterprises in other sectors (banking etc.) leasing bandwidth from telecommunications providers. In such an environment, we have the possibility of a number of parties owning resources within a single network which is both horizontally and vertically segmented. Each of these parties clearly has an incentive to see that overuse does not degrade network performance entirely but also an incentive to make the greatest possible use of their network ownership. Since these parties cannot agree each traffic policy decision individually, conflicting incentives must be reconciled outside the traffic management domain. Typically this is achieved through the setting of policy by sub-network owners within the remit of their own resources. The static nature of these policies and the conflict between them at sub-network interfaces cause institutionalized under-use of the network as a whole.

Given this background, the resource allocation problem in a network with multiple, non co-operating stakeholders can be recast as the problem of reconciling competition between self- interested, information-bounded agents. An effective mechanism for achieving this goal in the real world is the market economy. Therefore in this paper we present some preliminary work towards a telecommunication network management framework modelled on a market economy.

3   System Architecture

This section describes the top-level design of our market-based system. It describes the system architecture (section 3.1), the agent interaction (section 3.2) and the way in which resource commitments are handled (section 3.3).

3.1   A Layered System Architecture

We have designed our system as a three layered model (Figure 1). The lower layer represents a circuit-switched telecommunications network configurable with respect to the number and connectivity of its nodes and links. Resources on this layer are allocated by agents deployed at the nodes in the network which have access to real-time information about the state of utilization of links which are loaded from that node only. The middle layer is the agent-based network management system. It consists of three types of economic agent and two types of market institution which allocate resources in response to the buying and selling behaviour of the economic agents. The third layer is the user layer at which the system interfaces to the call request software. Within the call request software, call prioritization policy can be determined for different priorities of traffic at the user interface by setting spending limits for particular levels of call priority at the source nodes of the network.

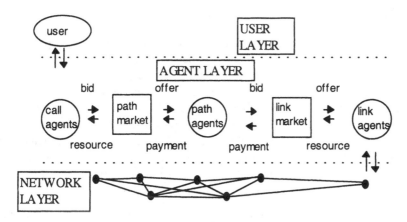

Figure 1. System Architecture Layered model of the relationship between the telecommunications network, the market-based multi-agent system and the user.

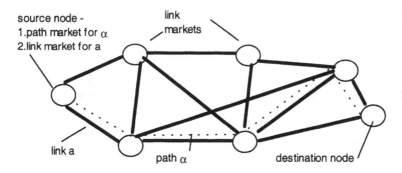

Figure 2. Network Topology Sample network configuration used for experiments.

In more detail: Link agents sell the underlying resources of the network (i.e. the capacity to carry a traffic along a link). Path agents buy resources from link agents which are then bundled into path resources which can carry traffic across the network. A path agent makes buying decisions within budget constraints to acquire link resources and then subsequently offers to sell the bundles of these resources (as paths) to caller agents which represent the end user. The system has an agent for each link and a market at the source node of that link serving path agents wishing to buy resources on that link. At the path level it has a path agent for each established path across the network (currently three for each source-destination pair) and a path market for each source destination pair in the network. Therefore in the example network shown in Figure 2. there are 24 link agents and link markets, 126 path agents and 42 path markets.

3.2   Agent Interactions

The agents communicate by means of a simple set of signals, which encapsulate offers to sell, bids to buy, commitments of and payments for resources. For convenience, we couple the resources and payments with the offers and bids respectively, this allows for a rapid protocol for resource and information exchange and is used because the time frame in which the resource allocation is carried out is critical to the responsiveness of the system. The speed of the negotiation and allocation protocol is traded off against its lack of flexibility when committing resources. This lack of flexibility may have a detrimental effect on allocative efficiency which in turn could contribute to the under performance of the market-based system across a range of network loadings. Analysis of this trade-off is part of ongoing work.

When a sale is agreed no further communication need take place before the resource can be used or offered for re-sale by the buyer. Buyers and sellers do not communicate directly with one another or amongst themselves. All interactions are by means the of institutions (markets). The market institutions also broadcast the clearing prices at which trades are agreed, so the agents have more information upon which to base their trading behaviour. (This has the effect of disseminating network level utilization information to all agents in the system).

The negotiation between buyer and seller agents is mediated by means of a market institution. The type of institution chosen for this purpose in our model was the double auction. A double auction is a market in which buyers and sellers place bids and offers simultaneously, the market matches the bids to the offers thereby setting the price and allocating the goods accordingly. The double auction was chosen for the following properties: It allows multiple buyers and sellers to negotiate simultaneously, it provides a dense set of market clearing price information and it allows supply and demand to be reconciled at the same time [4].

We have chosen to use a variant of the double auction which does not depend on the order or frequency of bids and offers arriving from the participating agents to allow for ultimate migration of the system to a distributed environment in which synchronous message delivery cannot be guaranteed. The variant of the double auction used in the current experiments works as follows: within a trading round, buyers and sellers submit bids and offers for known blocks or allocations of goods. After the auction stops accepting bids and offers (in this experiment agents can submit as many bids and offers as they wish in each time step), it ranks bids from highest to lowest and offers from lowest to highest. Then the highest bid and the lowest offer trade at a price midway between the two. This process continues until no offer remains at a price lower than a bid, at which point all bids and offers remaining are returned to their originators [5]. To make the clearing of resources and payments exchanged in this way efficient, we require that resources be coupled with offers and payments be coupled with bids.

The final role of our market institutions is to ensure acquaintance between agents that are interested in buying and selling particular resources. We distinguish two kinds of markets: one dedicated to mediating the exchange of link resources sold by link agents to competing buyer agents and the other dedicated to selling path resources by path agents to caller agents. These two markets are called the link market and the path market respectively (see Figure 1.).

3.3   Commitment and De-commitment of Resources

Once a price has been agreed between a buyer and seller (through the market institution), the seller's resource is committed to the buyer. The buyer must pay a rental for the use of this resource for as long as it remains allocated to it. In the event that a buyer cannot meet this payment, the resource is de-allocated and reverts to its original owner. A resource which is used profitably (re-sold at a profit) will never be de-allocated under this mechanism - meaning a call cannot be cut-off once it is connected. In addition to de-allocating resources that are not paid for, we also implement a mechanism for the bankruptcy of an agent which is holding resources but not selling them on to callers. Again we prevent calls from being cut-off under this rule by transferring profitable resources to the agent which replaces the bankrupt. Eventually the bankruptcy mechanism will be used to allow the system to learn differential loading patterns and network seasonalities.

4   Agent Design

This section describes the behaviour of the agents, paying particular attention to the functions which determine the prices at which they offer to buy and sell resources. The agents make their bids and offers so as to maximise their income if they are selling and to minimize their expenditure if they are buying. Path agents which both buy and sell resources, act so as to maximise their profit from ongoing transactions. Gode & Sunder [6] have shown that under certain conditions double auction type markets have the property of generating efficient allocation of resources among zero intelligence agents (i.e. those that do not reason with historical information to predict prices). This result led us to experiment with a relatively simple agent design to determine if routing decisions could be generated by a market mechanism in which agents used pricing functions based solely on endowments and the last known clearing price of their respective markets.

4.1 Link Agents

Link agents can be thought of as the producers of natural resources in the economy. As producers, they wish to maximise the income they derive by selling network resources to path agents. They do this by offering to sell at a relatively low cost when demand is low (in order to continue to acquire contracts) and competition between path agents is less, and at a higher price when demand is high (in order to exploit the greater competition among buyers). The price at which link agents offer resources is set in the following manner:

Let A be the set of link agents, and let $a \in A$ be a specific link agent. Each link agent (a) has a set of resources (R_a) which it can sell in the link market - in this case a unitary resource is sufficient (bandwidth) to carry one call across the link which that agent represents. R_a is composed of n equal bandwidth slices $\{r_{ai}, ...r_{an}\}$ which are allocated sequentially from the beginning.

Thus if r_{ap} ($1 \leq p \leq n$) is the next free resource then p-1 have already been allocated. After allocating r_{ap}, a will have n - p free resources. a's pricing function P for the next free resource is given by the following formula:

$$P(r_{ap}) = e^{(p/n)} \qquad (1)$$

This function was chosen so that once resources begin to be allocated on a link, the price rises quickly. This means that the marginal cost of obtaining the next resource (when the network is loaded) will be greater than that of obtaining the last.

4.2 Path Agents

Path agents act as both buyers of link resources and sellers of path resources. We detail their buying behaviour in section 4.2.1 and their selling behaviour in section 4.2.2. In general, however, path agents wish to buy resources cheaply from the link agents, and sell them on at a profit to the end-consumers. To do this, they bid competitively to acquire resources, which they then sell on to callers at a price not less than that paid for them. As a secondary factor the agent attempts to sell as many calls as possible (since these bring in revenue) while keeping inventory at a minimum (since this must be paid for).

4.2.1 Buying Behaviour

A path agent is actually constructed as a buying team which has a centralised budget for the whole path which it divides equally between all team members. Each member of the buying team places a bid to buy resources from a particular link agent at the market at which those link resources are sold.

Let α be a specific path of length n through the network and let PA_α be the path agent responsible for maintaining that path. PA_α will be composed of a team of n sub-agents ($PA_{\alpha,1}, \ldots PA_{\alpha,n}$) each of which is responsible for buying a particular link ($\alpha_i \rightarrow \alpha_{i+1}$) in the overall path.

Assume PA_α has a specific set budget B for buying its needed resources which it divides equally between its team of buying agents. Thus each PA_α has a budget of B/n. Given the selling price $P^t_{i \rightarrow i+1}$ of a path component at time t, the price bid by PA_{α_i} at the next time instant is given by the following formula:

$$P^{t+1}_{i \rightarrow i+1} = (P^t_{i \rightarrow i+1} + (B / n)) / 2 \qquad (2)$$

This function sets the price that the member of the buying team of the path agent offers for a link resources in the centre of the range between the last price paid (lower bound) and the budget for that link (upper bound). More sophisticated strategies are obviously possible.

4.2.2 Selling Behaviour

When a path agent has acquired sufficient resources, from each of its buyer team members, to carry a call across the whole path, it may offer a path resource (i.e. a bundle of an equal amount of resources from each of its buyer team members) for sale to callers. To remain profitable, the agent must not sell these resources at a price any lower than that at which they were acquired jointly by the buyer team. To maximize its profit the path agent should offer to sell this path resource at a price close to (but slightly below) that which it believes callers will be willing to pay. Additionally, the path agent should set its offer price to maximize throughput and minimize inventory.

To be able to offer a path α ($\alpha_1 \rightarrow \alpha_2 \rightarrow \ldots \rightarrow \alpha_n$), the path agent has to have bought resources for each of the path links ($\alpha_i \rightarrow \alpha_{i+1}$, for all $1 \leq i \leq n$). Let CO_α be the combined cost paid by PA_α for all the path's constituent components. In putting together the path α, PA_α may have acquired surplus resources which it stores in its inventory IN. PA_α may also have committed a number of path resources to calls already on the network, we refer to these resources as its throughput TH. The pricing function (see Equation 4. below) is adjusted by a function relating inventory and throughput to the offer price such that the agent offers resources for sale more cheaply when inventory is high and throughput is low. [1]

$$f(\text{IN, TH}) \rightarrow [0,1] \tag{3}$$

Using the price for which a path equivalent to α (i.e. with the same source and destination) was last sold at time t as a guide $P^t(\alpha)$, the price set at the next time instant P^{t+1} ranges between $P^t(\alpha)$ (upper bound) and CO_α (lower bound). (The agent will not offer resources for sale for less than the price for which they were purchased and cannot expect to have a price higher than that set by the auction met in the market). The offer price is set by the following formula:

$$P^{t+1}(\alpha) = CO_\alpha + f(\text{IN, TH}) \times (P^t(\alpha) - CO_\alpha) \tag{4}$$

At various times, path agents will find that they have bought unequal amounts of the resources they require. In addition to the cost paid to acquire these resources, path agents have to cover the ongoing cost of having these resources allocated to them. To avoid unnecessary costs, a path agent periodically examines its portfolio and returns surplus resources. In this case, surplus means those resources on a given link above the maximum resource allocation which that agent can provide across the whole of its path. More sophisticated mechanisms might have the agent allocate a disproportionate amount of money to acquiring resources to make up a shortfall and thereby balance their portfolio, rather than resort to the wasteful measure of discarding excess, or by making use of the concepts as levelled commitment [7] to allow flexible local deliberation.

4.3   Call Agents

The offers placed on the path market by path agents are allocated to the bids placed on that market by callers. Callers bid for path resources solely on the basis of the most recent price information they have. In future, it may be possible to have callers' bids determined by priority information from an external system (perhaps incorporating the amount of money that a real world caller is willing to pay and / or more sophisticated strategies based on price histories). For now, allowing callers to place their bids near

---

[1] In our experiments we used a number of functions to generate values in the range 0 to 1 from throughput and inventory state information but did not reach any firm conclusions about their effect on overall system performance.

to or slightly above the last price of which they are aware, enables us to successfully route calls across the network.

5 Experiments

To test our hypothesis that a market-based mechanism can effectively balance the traffic loaded onto a network, we conducted a number of experiments, at different network loads, to compare the performance of our system with that obtained using a conventional connection admission control process and static routing. It should be noted that our system does not discover previously unknown routes in the network, rather it seeks to make more efficient use of known routes by incorporating information about the cost to the network of using one route over another. Our simulator consists of a network of seven nodes irregularly connected by twenty-four directed links (see Figure 2). Traffic patterns are described by the total length of simulated time, the mean inter-arrival time between calls, and the mean call duration time. The inter-arrival times and call duration were determined using an exponential distribution. An equal probability of a call request between any source destination pair in the network was used, with approximately one in fifty calls requesting connection between any given source destination pair. We compare the performance of the two systems by measuring the number of calls accepted onto the network in a single time step. Since the total calls offered and mean call holding times are constant in each of our experiments this data can be used to calculate the intensity of traffic carried or lost by each mechanism using the standard telecommunication measure of traffic intensity (the Erlang).

In our simulator we use a negative exponential time distribution function to determine the interval between call arrivals and the duration of calls: Let $U(0, 1)$ be a random distribution function between 0 and 1. The inter-arrival time between calls and the call duration are calculated by the following formula:

$$f(x) = -\beta \ln U \qquad (5)$$

Where β is the desired cumulative mean inter-arrival time or call duration, respectively.

The inter-arrival time was varied between 1 and 0.1 seconds the latter representing a dense call arrival. Average call duration was varied between 50 seconds and 500 seconds. Both of which may be considered relatively short call durations. The simulation was allowed to run for 1000 seconds in each case. The interval shown in the graphs (300 - 350s) was chosen because it represents the transition point of loading for the intermediate traffic regime (i.e. the point at which most links in the network become saturated most of the time).

The experimental parameters were chosen to test the trial network using both traffic management mechanisms under a variety of network conditions. We tested the network using traffic ranging from light, in which a low call blocking rate would be expected (section 5.1), through intermediate (section 5.2) and heavy (section 5.3) in which most calls will not be admitted to the network.

5.1 Sparse Call Arrival – Short Call Duration

Both mechanisms perform reasonably well under this regime which represents a fairly light load for the network (Figure 3). As can be seen, the market-based mechanism drops more calls on average than the conventional static routing connection admission approach. This is in all probability due to opportunities for trade being missed due to inflexible trading strategies and negotiation protocols. It should be noted that the comparatively good performance of the static routing mechanism under light network loading is due to its sub-optimal decisions having a short duration relative to the inter call arrival time.

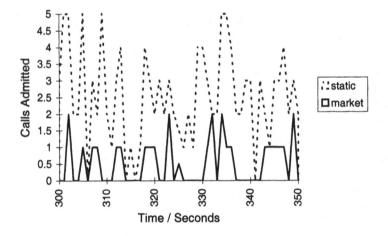

Figure 3. Call admission rate of static and market-based routing mechanisms under conditions of sparse call arrival and short average call duration.

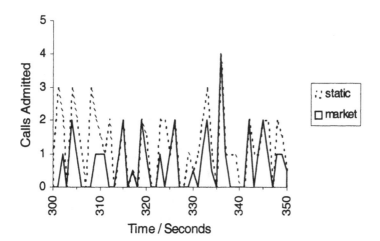

Figure 4. Call admission rate of static and market-based routing mechanisms under conditions of intermediate call arrival and intermediate average call duration.

5.2 Intermediate Call Arrival Medium Call Duration

These results show the static routing network at the transition point beyond which call loss becomes much more likely due to link saturation (Figure 4). The fact that static routing provides a relatively inflexible means of allocating calls to the network means that once network saturation occurs the ability to reroute traffic along alternative routes will not allow the network to make further use of its unused capacity to carry calls.

5.3 Dense Call Arrival Long Call Duration

The static routing mechanism does not factor in overall performance when allocating a call to a path and therefore the links of the network are allocated to the best paths under light network conditions. Under heavier conditions, these paths lead to further congestion.

At the point where the static routing mechanism is significantly past saturation point the market-based mechanism performs comparably in finding resources on the network to place calls (Figure 5). This is because market-based routing determines the cost of allocating link resources to paths and paths to calls such that it becomes very expensive to use a particular link past a certain point. At that point second and third best routes become cheaper alternatives and this reflects the underlying utilization of the resources throughout the network rather than at one part of it (the path in question).

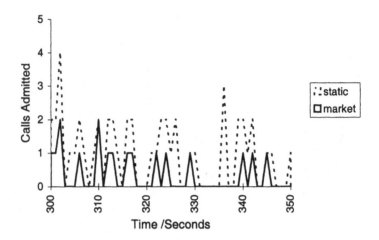

Figure 5. Call admission rate of static and market-based routing mechanisms under conditions of dense call arrival and longer average call duration.

5.4 Discussion

These results show that the ability of our market-based mechanism to make decisions in the presence of some information about the wider system state is of greatest value in situations of heavy network loading. However, under conditions of light and intermediate loading the market-based mechanism clearly under performed. This may be attributable to the course granularity of the resource units relative to the allocation problem being solved, or the fact that the relatively small number of agents meant that the competitive pressure of the market was not as great as it should have been. The simplistic method for de-allocation of resources may also be a contributing factor to the inefficiency of the market-based mechanism described in this work. There is also considerable room for improvement in the design of our agents themselves. The functions which they use to determine the prices to bid and offer for resources are fairly primitive and many opportunities for trade may be lost because of this.

In this work we make some simplifications about the operation of market-based control and static routing that do not properly reflect a real world situation. We allow markets to clear instantaneously and the connection admission decisions of the static routing mechanism to operate instantaneously. This is not sustainable in a real system in the case of static routing because call set up delay plays a significant role in (adversely) affecting the efficiency of the decisions made. It may be possible to overcome this problem in a market-based system by having some of the trading decisions which precede call set up being made before that call is requested through pro-active trading. The effect of time delays on the efficiency of market-based control mechanisms have been studied (particularly in the telecommunication domain) by Kuwabara, et al. [8] and in the domain of job sharing among networked computers by Chavez, Moukas & Maes [9]. Considerable work remains to be done in assessing the lag time of information passing around the system and the impact of this on allocative efficiency.

6   Related Work

Other authors have applied market-based control mechanisms to network control problems. Most notably Kuwabara, Ishida et al who have investigated an equilbratory approach (in which the market calculates the supply and demand volume equilibrium) to the allocation of resources in an ATM telecommunications network [8]. Our approach differs from theirs in that we ultimately wanted to explore the idea of having the resources needed to make a call across the network ready to be traded at the market at call set up time, thereby avoiding call set up overheads. This led us to treating each call as requiring a unique set of resources and therefore a departure from allocating resources according to a calculated demand-volume equilibrium. Wellman has also described an architecture in which different types of agents internalise the costs of a distributed multi-commodity flow problem by buying and selling resources at prices which reflect the cost to the network of routing traffic through shared links using local information [10]. Wellman's approach yields a balanced flow for the equilibrium state of the network in which the flows across each route are balanced by the market mechanism as predicted by network theory. However, in the domain we wished to investigate (a circuit switched network) each call is discrete and therefore could not be allocated to a number of routes simultaneously (bifurcated routing) as would be required to achieve equilibrium flows across competing routes.

7   Conclusions and Future Work

These preliminary results show the ability of a market-based mechanism to generate paths across a network and allocate calls to them dynamically in an environment in which network resources are owned by a number of different real world entities and therefore co-operation in network management domain cannot be taken for granted. Experimental investigation of the efficiency of this mechanism shows that this technique is most likely to be beneficial when the network is heavily loaded.

Our analysis of the results so far indicates that a more thorough treatment of the relationship between competitive pressures and market efficiency may lead to higher allocative efficiency in the market, and therefore smoother load balancing in the system. Therefore we intend to review how the pricing strategies of the agents are designed and the complexity of the functions required to achieve competitiveness. If the objective were to design the agents for profitability in harsher economic conditions than those simulated here (in our system callers continue to meet the prices asked by path agents no matter how high they inflate) much could be done to make the path agents more aggressive traders. For example, they could be given the ability to reason about the market over time to adjust their strategies accordingly. An alternative approach which we intend to investigate which makes use of speculating agents to stabilize markets has been shown to provide a worthwhile improvement in allocative efficiency [11].

Another interesting area for future research is the possibility that the market mechanism may allow for the resources needed to make calls to be allocated proactively. This would provide an advantage over conventional call set up / connection admission proceedures in that it may be possible to overcome the problem of call set up delay affecting the efficiency of the decision taken. The call set up properties of market-based and conventional routing mechanisms will be investigated in future work.

This work takes place within the wider context of investigation into the application of market-based control to various aspects of telecommunications network management. It is hoped eventually to extend the approach to design a network management system for connection oriented packet switched technologies such as asynchronous transfer mode (ATM) telecommunications networks.

Acknowledgements

This work was carried out under Engineering and Physical Sciences Research Council (EPSRC) grant No.GR/ L04801

8 References

1. Schwartz M. (1988) Telecommunications Networks - Protocols, Modeling and Analysis, (Addison Wesley. Publishing Company)

2. Appleby S. & Steward S. (1994) Mobile Software Agents for Control in Telecommunications Networks, BT Journal of Technology 12 - 2, pp 104 113

3. Schoonderwoerd, R., Holland, O.E., Bruten, J.L. (1997) Ant-like agents for load balancing in telecommunications networks,. in Proceedings of the First International Conference on Autonomous Agents. Marina Del Ray, California, 1997.ACM Press.

4. Friedman, D. & Rust J. (1991) The Double Auction Market: Institutions, Theories, and Evidence, Proceedings of the Workshop on Double Auction Markets held June, 1991, Santa Fe, New Mexico, Addison-Wesley Publishing Company

5. Davis D. D. & Holt C.A. (1993) Experimental Economics, Princeton University Press

6. Gode D.K. & Sunder S. (1993) Allocative Efficiency of Markets with Zero-Intelligence Traders: Markets as a Partial Substitute for Individual Rationality, Journal of Political Economy Vol 101-1 1993.

7. Sandholm T. (1995) Issues in Automated Negotiation and Electronic Commerce: Extending the Contract Net Framework, First International Conference on Multiagent Systems (ICMAS-95), San Fransisco, pp. 328-335.

8. Kuwabara K., Ishida T., Nishibe Y. & Tatsuya S. (1996) An Equilibratory Approach for Distributed Resource Allocation and its Applications to Communication Network Control, in Market - Based Control, A paradigm for distributed resource allocation, Ed. Clearwater S., World Scientific Publishing, 1996, pp 53 - 73

9. Chavez A., Moukas A., Maes P. (1997) Challenger: A Multi-Agent System for Distributed Resource Allocation, in Proceedings of the First International Conference on Autonomous Agents, Marina Del Ray, California, 1997. ACM Press.

10. Wellman M.P, A Market Oriented Programming Environment and its Application to Distributed Multicommodity Flow Problems; Journal of Artificial Intelligence Research 1 (1993) 1 -23.

11. Steiglitz K., Honig M. L. & Cohen L. M. A. (1996) Computational Market Model Based on Individual Action, in Market - Based Control, A paradigm for distributed resource allocation, Ed. Clearwater S., World Scientific Publishing, 1996, pp 1-27.

The Application of Intelligent and Mobile Agents to the Management of Software Problems in Telecommunications

Stephen Corley[1], Diego Magro[2], Fabio Malabocchia[2], Jens Meinköhn[3]
Luisella Sisto[2], Sahin Albayrak[4], Alexander Grosse[4]

1 BT Laboratories, Martlesham Heath, Ipswich, UK. IP5 3RE.
2 CSELT, Via G. Reiss Romoli 274 I-10148, Torino, Italy.
3 Deutsche Telekom, Berkom GmbH, Goslarer Ufer 35, D-10589 Berlin, Germany.
4 DAI Lab, TU Berlin, Sekr. FR 6-7, Franklinstr.28/29, 10587 Berlin, Germany.

Abstract. Agent technology promises to increase the flexibility and power of telecommunications management systems and services. This paper describes ongoing work that, through prototyping and experimentation, is aiming to understand the practical implications and benefits of applying agent technology to the management of problems related to service application software. Agents are seen to play a central role in enhancing flexibility, user friendliness and productivity, both for the end users and the support engineers. The agents can take care of much of the diagnosis, solution logistics and difficult software set-up procedures.

1   Introduction

In this competitive telecommunications world the drive for new and sophisticated services is fundamental. However, to survive, the efficiency of the management systems supporting the services and the quality of support perceived by the end users are paramount. The OMG [1] and TINA [2] architectures have made significant progress towards supporting a modular and flexible approach to control and management software. This will go some way to meeting the demands of modern telecommunications. However, the drive for additional techniques continues in order to attain an even higher degree of software flexibility and dynamic configurability in systems and services. Intelligent and mobile agent technologies promise to enable this level of capability.

This paper describes part of the work of the EURESCOM[1] (European Institute for Research and Strategic Studies in Telecommunications) Project P712 [3]. This project

---

[1] EURESCOM is a German Private company owned by 23 Shareholders from 22 European countries, whose overall mission is to carry out pre-competitive R&D Projects (see http://www.eurescom.de/).

has the main objective of assessing the maturity, benefits and practical implications of using intelligent and mobile agent technology for building service and network management systems. Service and network management are broad terms covering major functional activities concerned with operating a telecommunications business. Many functions are required to provide service offerings to customers and to operate the supporting networks. The functions can be broken down into the general areas of fault, performance, security, accounting and configuration management. Service management is customer facing and includes the function of customer care. The project has chosen two case studies in the domain of service and network management as the basis for its investigations. Through prototyping and experimentation valuable experience and information will be generated which will be extremely useful for anyone considering the deployment of agent technology in their business.

This paper describes one of the case studies; that of software problem management which includes aspects of the fault, configuration and customer care management areas. The other case study, dynamic configuration for co-operative working is described in [4].

The remainder of this paper is structured as follows: the description of the software problem management scenario is provided in section 2. The main part of this paper, section 3, is devoted to the realisation using agents. The architecture and the agents involved are described in detail. Section 4 discusses some of the issues arising from the work so far. Finally, a summary is given in section 5.

2   The Software Problem Management Scenario

In the world of telecommunications today, new services are appearing at an ever increasing rate, fuelled by both 'technology push' and 'market pull'. The maintenance of these services relies on network maintenance as far as the reliability of the physical processing and transmission infrastructure is concerned. However, the services themselves are generally implemented via special purpose pieces of software that form a complex system. The maintenance of such systems involves not only fault diagnosis and bug repair, but also their extension with other software modules or the update of one or more components. Due to the increasing diversity and complexity of services, end users (customers) are likely to experience difficulties when interacting with the service. This added factor means that service maintenance becomes a specialisation of the more general class of activity of software problem management.

As described in the introduction, the objective is to understand, by implementing and experimenting with a prototype, the benefits and practicalities of applying agent technology to the management of software problems in the context of telecommunications services. However, this is a very wide area and so to narrow the scope the electronic mail service has been selected as a focus.

In this domain, various problems can be experienced that require different solutions: a common one is, for example, the inability of reading[2] a mail content or an attachment. This could be due to some software incompatibilities between the sender and the receiver or, more specifically, to the unavailability at the receiver side of the appropriate content handler. In this case, the alternative solutions could be as follows:

- install in the receiving computer the appropriate content handler,
- use an appropriate converter at the receiving end,
- request the sender to re-send the mail with another format, which may require the use of a converter.

Other email problems could include router configuration problems, oversized files, queuing problems, security restrictions and list maintenance. The content handling problem has been taken as the first type of problem to be solved in the case study and is used as the basis of discussion throughout the remainder of the paper.

In the presence of a software problem, we are focusing on two broad approaches to its resolution:

1. *automated maintenance:* the service maintenance system attempts, without any human involvement, to solve the problem completely or with a temporary graceful degradation in service performance. Ideally, the user should not become aware that a problem has occurred. In the context of the email service, the scenario could be that the system detects that the user wouldn't be able to see (or hear) the message content so the system itself searches for the appropriate content handler and it installs it in the receiving computer. If the search or the installation fails, a message can automatically be sent to the sender to ask for the message to be re-sent in another format. In this way the receiving user wouldn't be aware that an unreadable mail was sent.

2. *effective support for the expert technician:* if the system fails in automatically solving the problem, the failure will become rapidly apparent to the user who will contact an expert technician to provide a solution. Ideally, if the maintenance system was aware of the problem then the technician would have already been notified before the user made contact. On notification of a problem, by whatever means, the technician is supported by the system in the diagnosis and repair activities through the system making knowledge and experience relevant to the problem at hand readily available. The system will also support the resolution of exceptional problems requiring collaboration among more than one technician by ensuring that the most appropriate technicians are put in contact.

In the first case, the benefit is that tasks normally solved manually can be automated, reducing the time and human effort required to solve the problem. By automating as much as possible the technicians can make more effective use of their skills and time.

---

[2] We say 'reading', but, actually, the mail can contain data in various forms: text, voice, video, etc.

In the second case, the system offers support to the diagnosis and the repair activities leading to more efficient use of the technician's time. In both cases the quality of service as perceived by the users is improved.

3 Realisation Using Agents

3.1 Prototype Architecture

The prototype is based on the use of agent technology both in those capabilities concerning the automated maintenance and in those ones relevant to the effective support for the expert.

The diagram in Figure 1 summarises the architecture of the overall prototype.

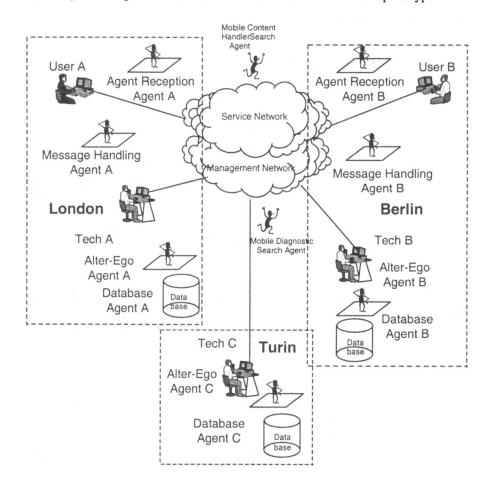

Fig. 1. Prototype Architecture

Without loss of generality, we assume that the territory is organised around three sites (London, Berlin and Turin) and that in each one there is one user of the mail service and one technician. The Turin user does not play a role in our scenarios and is therefore not shown. The maintenance agents relevant to the user are two static agents, the Agent Reception Agent (ARA) and the Message Handling Agent (MHA), and one mobile agent, the Content Handler Search Agent (CHSA). On the technician side, we have two other static agents, the Alter Ego Agent (AEA) and the Database Agent (DBA), and one mobile agent, the Diagnostic Search Agent (DSA). In each site, there is a database containing the problem solving past episodes that the local technician successfully completed. In this way, the service maintenance infrastructure possesses a distributed case memory.

For automated maintenance the user agents play the most important role, while the technician agents are those that offer support to the expert. In addition, the technician agents can also support the user agents in the problem resolution process so creating a society of co-operating agents.

3.2 Description of Agents Involved

The roles of each type of agent are described below.

Message Handling Agent (MHA)
Each user will have a static agent responsible for handling the messages received and sent by the user. This agent will have the capability to detect and diagnose problems with messages on behalf of the user. Although this agent is static, it may co-ordinate the use of smaller mobile agents (Content Handler Search Agents) to perform its tasks. The agent may communicate with the Alter-Ego Agent to make use of its diagnostic capability. In the event of problems which it cannot handle itself, it will flag these up to the user so that they can notify first line support manually.

Agent Reception Agent (ARA)
This agent is responsible for communicating with and controlling incoming mobile agents (owned or alien) which may wish to 'land' on the site. It is effectively a 'gatekeeper' to the local machine and handles the security, registration, de-registration tasks and controls access to local resources.

Content Handler Search Agents (CHSA)
This agent is mobile as it must be able to travel over a network to search for an appropriate content handler. It must carry with it a description of the requirements and be able to interact with agents at remote sites in order to obtain the handler software (this may involve negotiation over costs, time-scales and licensing agreements). It must also be able to install the software on the local environment.

Diagnostic Search Agents (DSA)

This agent is mobile as it must be able to travel over a network interconnecting the sites and access the required diagnostic information. Hosts or sites may be temporarily down preventing the agent from landing on a site, but as the task does not have real time requirements, the time-out for getting the answers back from the network can be reasonably set in the order of hours.

Local Database Agent (DBA)

This agent is static and there is one located at each site. It has the responsibility for receiving the problem description from the incoming search agent (DSA), searching over the local database and passing back to the DSA a list of the matches found ranked according to the degree of relevance.

Alter-Ego Agent for the technician (AEA)

This agent is static and has the most complex capabilities of all the agents within the scenario. It has responsibility for retrieving the task from the technician, dispatching the mobile agents (DSAs) to search for the required information, collating the received answers and presenting a coherent picture of the possible resolutions to the technician. One additional task that this agent may be called to perform is to set up a phone meeting in the case of a match being found which requires further clarification or discussion with another technician. This is performed when the local technician agrees to a phone-based consultation and requires that both agents know when their technicians are available.

3.3 The Problem Management Process

The MHA has the main role of monitoring all incoming mails and if it discovers that there is a problem, such as the user will not be able to 'read' a particular mail if they open it, then the MHA tries to solve the problem. To achieve the goal of finding the solution, it first establishes the characteristics of the problem, which in this case would include the type of the unreadable content and using its reasoning capabilities determine an appropriate solution or set of possible solutions. If necessary, the MHA will call upon the AEA to provide a diagnosis service. In this case, the MHA requests an <u>exact</u> solution to the problem. Assuming the solution is determined to be to find and install a new content handler, then the MHA creates one or more CHSAs to find the specified content handler. The CHSAs are mobile, so they travel to the destination site and perform the search locally. At each site there is one ARA, which in addition to being responsible for receiving and trusting the incoming CHSAs, is also responsible for the content handlers available from that site. If the required content handler is available, the CHSA and ARA will enter into a negotiation over terms and conditions for its use. CHSAs which fail to find anything suitable destroy themselves. The returning successful CHSAs report back to the MHA which chooses one and asks it to install the content handler.

If no CHSA comes back or if the installation fails, the MHA can try another solution such as asking the sender's MHA to re-send the mail in an alternative specified format. If this fails (normally because it is unable to comply with the requested content type) and the MHA has no other options, or indeed, it was unable to find a solution in the first place then the receiver's MHA notifies the problem to the user, who can ask the technician for a manual solution. The MHA can also notify the AEA direct.

When the problem must be handled by a technician, the technician can benefit from the maintenance infrastructure support both in finding past solutions to identical or similar problems and in organising a phone meeting with a colleague. If the technician can't solve the problem alone, then the local AEA can be called upon to try to find a solution to a similar problem that someone has solved in the past. This goal is achieved through a simple form of co-operative case-based reasoning [5], [6]. The AEA firstly asks the local DBA to retrieve from the local case memory a stored similar case. The DBA is the agent responsible for retrieving and storing the cases from and in its case memory. Since in a real situation the cases stored can contain some parts in free-text form then natural language matching could be used [7]. If no case is found, then the AEA broadcasts one DSA to each other site. The DSA is a mobile agent that carries with it the problem description and that lands on the destination site and asks the DBA in that site to perform a lookup. If some cases are found, they are ranked and the best ones are carried back by the DSA to the original site, otherwise the DSA destroys itself. The returning DSAs communicate to the AEA the results of the search and then they die. If no DSA comes back, it means that nobody has already solved a similar problem, so when the technician eventually solves it, the problem description along with its solution is stored in the local case memory. If some cases have been found, the AEA chooses the best ones and it presents them to the technician.

The similarity degree between the problem and the retrieved cases can be of three types: perfect, good or slight. The first two match degrees indicate that the retrieved cases contain enough information for the expert to solve the problem, while a slight match means that the technician may need help to adapt the retrieved solution to the problem at hand. The AEA, in this case, asks the AEA resident in the site at which the case with the slight match was found to arrange a telephone meeting between the two technicians to allow them to team up to solve the problem. The AEAs know the diary of their technicians and have a set of rules to manage it in an intelligent way. If the AEAs can't find an appropriate meeting time for both the technicians, no meeting can be set up. In this case, or the case where no match was found, then escalation to the next line of support, which could be the developers of the application software, will be required. Once solved, a problem with only slight matching cases in the existing case memory is stored in the local database.

3.4 Implementation

The partners involved in the case study are Deutsche Telekom Berkom (supported by
TU Berlin), BT and CSELT. Figure 2 shows the major physical blocks associated
with the prototype implementation platform and how the agents are distributed among
the partners.

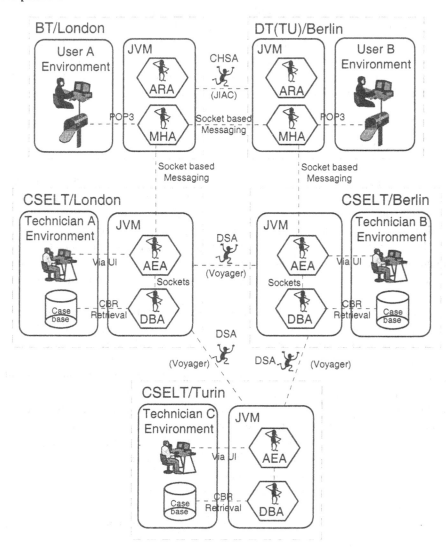

Fig. 2. Prototype Implementation Platform

The implementation platform for the prototype combines a number of Java based
development environments. These include Voyager [8] from ObjectSpace for

providing mobility in agents, JESS (Java Expert System Shell) [9] for reasoning capabilities and a number of environments developed in-house by the project partners, for example, Java Intelligent Agent Componentware (JIAC) from DT is used as second environment for providing mobile agent capabilities. The agents execute in a Java Virtual Machine (JVM) on each host. The Internet is used as the transport layer for inter-agent communications and the migration of mobile agents.

CSELT is responsible for implementing the agents which support the expert technician. BT and Deutsche Telekom are responsible for the automated resolution agents. The prototype development is due to be completed by August 1998. Once complete a number of experiments will be carried out to gather data on particular metrics and properties of the agent-based solution. Some of these metrics and properties are mentioned in the discussion below (section 4).

4 Discussion

Apart from software problem management being of vital importance to the telecommunications industry, it provides an ideal context for investigating agents in a very wide range of situations, such as, interaction with human users, automated problem solving, co-operation among agents, negotiation about contracts, interaction with legacy systems, integration with traditional AI techniques and interaction with foreign host resources. All of these situations are likely to appear in many other domains and therefore the knowledge gained in this study will be quite generally applicable.

The agents used in this investigation exhibit the important characteristics of agents; autonomy, intelligence, mobility and social abilities, to varying degrees. The MHA is strong in autonomy which is natural for a piece of software that must handle problems without the explicit request of a user. The MHA also has to have a certain degree of intelligence since it must know and reason about the domain of email problems. It is not necessary for the MHA to have inherently very strong diagnostic capabilities; it makes up for this by its ability to co-operate with other agents to get the problem solved. This co-operation is a key feature of multi-agent systems.

The AEA is similar to the MHA in its characteristics except that it is stronger in its intelligence, due to its diagnostic capabilities, and also in its social abilities, due to its requirement to interact and support human users.

Mobility, of course, is the predominant characteristic of the DSA and CHSA agents. They are autonomous in the sense that once they have been launched on a search they are, barring intervention from malicious agents, under control of their own destiny. Indeed they choose themselves whether they 'live' or 'die'. It is interesting to consider how much intelligence or social ability a mobile agent such as these can have. Intelligence and high-level communicative abilities required for flexible co-operation is demanding on computing resources. The limited bandwidth, processing power and storage of current computing infrastructures means that the footprints of mobile agents are necessarily constrained. One of the areas of experimentation in the case study is to assess the level of complexity that is currently practical for mobile agents to possess.

Finally, the DBA is hardly an agent at all. It is simply a database (case base) search engine with a wrapper to provide high-level communicative abilities via an agent communication language.

Learning is often quoted as a important capability of intelligent agents. It is interesting to note that although none of the agents inherently having any learning capabilities, the system overall does improve with use and over time as new cases are added. It is the human technicians which are the source of new knowledge; the agent-based maintenance system simply stores and retrieves it in an efficient way. Of course, the agent-based systems would be ideal for the inclusion of machine learning techniques, but this is outside the scope of the current investigations.

The use of mobile agents for searching has a number of advantages over traditional Remote Procedure Call (RPC) approaches. The common advantage often quoted is that of reduction in network traffic due to only retrieving the data required rather than all of it only to throw all but a few items away. This is valid for our scenarios too, but in addition, we have agents with the ability to negotiate. This negotiation, if carried out by static agents could create a significant unwanted traffic overhead. The mobile agent approach also allows parallelism by sending out multiple agents simultaneously, and load balancing by off-loading activities onto other hosts. Both of these features are useful in problem management, particularly where service contracts mean time may be short or where the management system itself is experiencing problems (e.g. faults or unexpectedly high throughput).

It is worth saying a few words about the case-based reasoning performed. In supporting the expert activity, the system only performs a case retrieval in a distributed case memory, while the application and the possible adaptation of the retrieved solution are done by the technician. This is particularly relevant in the case of automated maintenance; the MHA can ask for a past solution and it can apply it, but it must request an exact match. In this way it can apply (or attempt to apply) the retrieved solution without any adaptation: the type of CBR performed is, thus, a *precedent case based reasoning*. An obvious extension is a management system that can perform also the adaptation phase, and thus have a *case based problem solving* capability.

Until now we have implicitly assumed that agents know each other and that they also know the services that each one can offer. However, this may not be a realistic assumption in a real-world solution, except for small, fixed and closed systems. Indeed, in this situation you could argue that you would be better off NOT using agents. So, if we want to build an 'open' system consisting of a society of agents, in order to allow a more flexible co-operation among the agents and the dynamic addition of new types of agents, some questions arise: how can the agents discover each other? How can they know the services offered by each one? How can they communicate among themselves? Some answers to these questions may come from the standardisation efforts that the agent community is progressing (e.g. FIPA97 [10]). It is essential to understand if the overhead of defining a system compliant to some standard is counterbalanced by the benefits of the 'world openness'. The project will report on this issue.

A crucial issue inherent in this, and indeed any, system using mobile agents is that of security. Inevitably there will some individuals who view mobile agents as yet another means of causing havoc in computer systems. Even mobile agents with non-malicious intents could cause disruptions in networks under unexpected circumstances. Although recognised as crucial, due to limited resources, a full analysis is outside the scope of this study. However, the opportunity will be taken to understand more about the size and implications of the security problem.

5 Summary

Modern telecommunications is characterised by continuous change with distributed, often mobile, data and control. Agent technology promises to make a dynamic world such as this manageable. Few, if any, agent-based management systems have been deployed in large scale industrial settings. Although agent technology is maturing rapidly, what has not been proven are the full practical implications of such an approach, in terms of costs, performance and technical feasibility. This paper has given an overview of the software problem management study currently in progress within EURESCOM project P712. Through prototyping and experimentation this study will provide a deep understanding of the use of a range of agent types (including intelligent, mobile and co-operating agents) for managing problems in the context of telecommunications services. The agent-based architecture has been described with a detailed description of the agents involved. An overview of the physical implementation platform has been given. Some of the issues arising out of the study so far have been discussed. A future paper, expected in 1999, will detail the full findings of the study once the prototype and experiments have been completed.

Acknowledgements

The authors would like to thank the EURESCOM organisation, particularly Juan Siles, and other members of the Project team for their support and help in producing this paper.

6 Disclaimer

This document is based on results achieved in a EURESCOM Project. It is not a document approved by EURESCOM, and may not reflect the technical position of all the EURESCOM Shareholders. The contents and the specifications given in this document may be subject to further changes without prior notification. Neither the Project participants nor EURESCOM warrant that the information contained in the report is capable of use, or that use of the information is free from risk, and accept neither liability for loss or damage suffered by any person using this information nor for any damage which may be caused by the modification of a specification. This document contains material which is the copyright of some EURESCOM Project Participants and may not be reproduced or copied without permission. The commercial use of any information contained in this document may require a license from the proprietor of that information.

7 References

1. Object Management Group, "http://www.omg.org/".
2. Telecommunications Information Network Architecture, "http://www.tinac.com".
3. Corley, S., Tesselaar, M., Cooley, J., Meinköhn, J., Malabocchia, F., Garijo, F., "The Application of Intelligent and Mobile Agents to Network and Service Management", Proceedings of IS&N98, 1998.
4. Garijo, F., Tous, J., Corley, S., Tesselaar, M., "Development of a Multi-Agent System for Cooperative Work with Network Negotiation Facilities", Proceedings of IATA'98, 1998.
5. Aamodt, A., Plaze, E., "Case-Based Reasoning: Foundational Issues, Methodological Variations, and System Approaches", AI Communications, Vol. 7, No. 1., pp. 39-59, 1994.
6. Plaza, E., Arcos, J.L., Martin, F., "Cooperative Case-Based Reasoning", in G. Weiss (ed.) "Distributed Artificial Intelligence meets Machine Learning", LNAI, Springer Verlag, n. 1221, pp. 180-201, 1997.
7. Watson, I., Watson, H., "Case-Based Content Navigation", in Macintosh, A., Milne, R., (Eds.), Proceedings of ES97, pp. 185-195. (1997).
8. Voyager, "http://www.objectspace.com/voyager/".
9. JESS, Java Expert System Shell, "http://herzberg.ca.sandia.gov/jess".
10. FIPA97 Specifications, "http://drogo.cselt.stet.it/fipa/".

Distributed Fault Location in Networks Using Mobile Agents

Tony White, Andrzej Bieszczad, Bernard Pagurek

Systems and Computer Engineering, Carleton University,
1125 Colonel By Drive, Ottawa, Ontario, Canada K1S 5B6
email: {tony, andrzej, bernie}@sce.carleton.ca

Abstract. This paper describes how multiple interacting swarms of adaptive mobile agents can be used to locate faults in networks. The paper proposes the use of distributed problem solving using mobile agents for fault finding in order to address the issues of client/server approaches to network management and control, such as scalability and the difficulties associated with maintaining an accurate view of the network. The paper uses a recently described architectural description for an agent that is biologically inspired and proposes chemical interaction as the principal mechanism for inter-swarm communication. Agents have behavior that is inspired by the foraging activities of ants, with each agent capable of simple actions; global knowledge is not assumed. The creation of chemical trails is proposed as the primary mechanism used in distributed problem solving arising from the self-organization of swarms of agents. Fault location is achieved as a consequence of agents moving through the network, sensing, acting upon sensed information, and subsequently modifying the chemical environment that they inhabit. Elements of a mobile code framework that is being used to support this research, and the mechanisms used for agent mobility within the network environment, are described.

1 Introduction

The telecommunication networks that are in service today are usually conglomerates of heterogeneous, very often incompatible, multi-vendor environments. Management of such networks is a nightmare for a network operator who has to deal with the proliferation of human-machine interfaces and interoperability problems. Network management is operator-intensive with many tasks that need considerable human involvement. Legacy network management systems are very strongly rooted in the client/server model of distributed systems. This model applies to both IETF [1] and OSI [2] standards. In the client/server model, there are many agents providing access to network components and considerably fewer managers that communicate with the agents using specialized protocols such as SNMP or CMIP. The agents are providers (servers) of data to analyzing facilities centered on managers. Very often a manager has to access several agents before any intelligent conclusions can be inferred and presented to human operators. The process often involves substantial data transmission between manager and agent that can add a considerable strain on the throughput of the network. The concept of *delegation of authority* has been proposed [3] to address this issue. Delegation techniques require an appropriate infrastructure that provides a homogeneous execution environment for delegated tasks. One approach to the problem is SNMPscript [4]. However, SNMPscript has serious restrictions related to its limited expression as a programming language and to the limited area of its applicability (SNMP only). Although *delegation* is quite a general idea, the static nature of management agents still leaves considerable control responsibility in the domain of the manager. Legacy network management systems tend to be monolithic, making them hard to maintain and requiring substantial software and hardware computing resources. Such systems also experience problems with the synchronization of their databases and the actual state of the network. Although the synchronization problem can (potentially) be reduced in severity by increasing the frequency of updates or polling, this can only be achieved with further severe consequences on the performance of the system and the network.

An emerging technology that provides the basis for addressing problems with legacy management systems is network computing based on Java. Java can be considered a technology rather than merely as another programming language as a result of its 'standard' implementation that includes a rich class hierarchy for communication in TCP/IP networks and a network management infrastructure. Java incorporates facilities to implement innovative management techniques based on mobile code [5]. Using this technology and these techniques it is possible to address many interoperability issues and work towards plug-and-play networks by applying autonomous mobile agents that can take care of many aspects of configuring and maintaining networks. For example, code distribution and extensibility techniques keep the maintainability of networks and their management facilities under control. The data throughput problem can be addressed by delegation of authority from managers to mobile agents[1] where these agents are able to analyze data locally without the need for any transmission to a central manager. We can limit the use of processing resources on network components through adaptive, periodic execution of certain tasks by visiting agents. The goal is to reduce, and ultimately remove, the need for transmission of a large number of alarms from the network to a central network manager. In other words, our research focuses on proactive rather than reactive management of the network.

While Java technology provides a device independent agent execution environment, the use of mobile code in Network Management and the use of groups of agents in particular, generate a number of issues which must be addressed. First, how is communication between agents achieved? Second what principles guide the migration patterns of agents or groups of agents moving in the network. Finally, how are groups of agents organized in order to solve network-related problems? These questions motivate the research reported in this paper.

---

[1] The terms "mobile agent" and "mobile code" will be used interchangeably throughout this paper.

The remainder of this paper is organized in the following way. First, we briefly describe an infrastructure for mobile code that has been designed and implemented in Java. A mobile code taxonomy is then presented. The essential principles of Swarm Intelligence (SI) and, in particular, how an understanding of the foraging behaviors of ants [6] has led to new approaches to control and management in telecommunications networks are then reviewed. An agent architecture utilizing mobile code for the localization of network faults is then provided, along with an example of its use in a network scenario. The paper then concludes with a review of key messages provided and a review of planned future activities.

2 Mobile Code Environment (MCE)

A homogeneous execution environment for mobile code is considered extremely advantageous for the agent-based management of heterogeneous networks. Typically, an MCE contains the following components [7]: a mobile code daemon, a migration facility, an interface to managed resources, a communication facility, and a security facility.

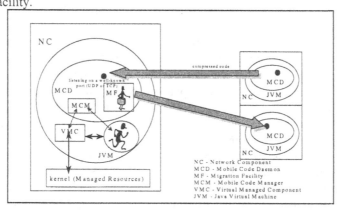

Fig. 1. MCE Components

It is assumed that a mobile code daemon (MCD) runs within a Java virtual machine on each network component (Figure 1). The mobile code daemon receives digitally signed mobile agents and performs authentication checks on them before allowing them to run on the network component. While resident on the network component, mobile agents access managed resources via the virtual managed component (VMC). The VMC provides get, set, event and notification facilities with an access control list mechanism being used to enforce security. VMCs are designed to contain managed information base (MIB) and vendor-related information. A migration facility (MF) provides transport from one network component (NC) to another. The mobile code manager (MCM) manages the agent lifecycle while present on the NC. For more detailed information on the MCE see [7].

Mobile code environments are connected with default migration patterns in order to form mobile code regions [9] with gateways between them. The migration facility is used to move a mobile agent from one network component to another, either within the same region or between regions. A single mobile code region will be assumed for the remainder of this paper. Individual mobile agents may use the default migration destination or use other algorithms in their choice of migration destination. Migration algorithms are presented in the section on agent architecture.

3   Mobile Agents Types

The management of networks using delegation and mobile code has seen the development of a taxonomy of agents [8]. Three principal types of mobile agents are defined. They are *servlets*, *deglets* and *netlets*. Servlets are extensions or upgrades to servers that stay resident as integral parts of those servers. Mobile agents constituting servlets are sent from one component to another and are installed as code extensions at the destination component; i.e. the agent typically migrates no further. For example, a servlet encapsulating the telnet protocol might be sent from one component to another in order to facilitate telnet access to the receiving component. Deglets are mobile agents that are delegated to perform a specific task and generally migrate within a limited region of the network for a short period of time, e.g. to undertake a provisioning activity on a network component. Netlets are mobile agents that provide predefined functionality on a permanent basis and circulate within the network continuously. An example of a netlet might be a component or service discovery agent or an agent constituting part of a distributed expert system. This latter example will be the subject of a later section.

In the management of networks using mobile code, the traditional client/server interaction represented by an SNMP agent reporting to a single workstation is replaced by a set of mobile agents injected by a management workstation that circulate throughout the network (typically) reporting only anomalous conditions found.

4   Swarm Intelligence

While the MCE enables the transfer of code from one component in the network to another and the principle of delegation a reason to use it, it does not provide for distributed problem solving by groups or societies of agents. This is the nature of Swarm Intelligence.

Swarm Intelligence [10] is a property of systems of unintelligent agents of limited individual capabilities exhibiting collectively intelligent behavior. An agent in this definition represents an entity capable of sensing its environment and undertaking simple processing of environmental observations in order to perform an action chosen from those available to it. These actions include modification of the environment in which the agent operates. Intelligent behavior frequently arises through indirect communication between the agents, this being the principle of stigmergy [11]. It should be stressed, however, that the individual agents have no explicit problem solving knowledge and intelligent behavior arises because of the actions of societies of agents.

Two forms of stigmergy have been described. *Sematectonic* stigmergy involves a change in the physical characteristics of the environment. Ant nest building is an example of this form of communication in that an ant observes a structure developing and adds to it. The second form of stigmergy is *sign-based*. Here, something is deposited in the environment that makes no direct contribution to the task being undertaken but is used to influence subsequent task related behavior.

Sign-based stigmergy is used in the foraging behavior of ants. The use of ant foraging behavior as a metaphor for a problem-solving technique is generally attributed to Dorigo [12]. It is considered central to our work. To date, three applications of the ant metaphor in the telecommunications domain have been documented [13], [14] and [15]. [14] embraces routing in the circuit switched networks while [15] deal with packet switched networks. Both [14] and [15] propose the control plane as the domain in which their systems would most likely operate. [14], in particular, provide compelling experimental evidence as to the utility of ant search in network routing.

5   Service Dependency Modeling

In order to drive the problem solving process -- that of fault finding -- a model of faults, or a concept of services and dependencies between them, is required.

Within the context of this paper, a network is said to provide services; e.g. private virtual circuits (PVCs). When a service is instantiated; e.g. a new PVC is created, it consumes resources in that network and subsequently depends upon the continued operation of those resources in order for the service to be viable. From a fault finding perspective, a service can then be defined in the following way:

$$S \mapsto \{(R_i, p_i)\} \tag{1}$$

where S is the service, R_i is the ith resource used in the service, p_i is the probability with which the ith resource is used by that service and the relational operator means depends upon. A resource R_i might be a node, link or other service.

For example, a PVC that spans part of a network might depend upon the operation of several nodes and T1 links. The links, in turn, might depend upon the correct operation of several T3 links that carry them in a multi-layer virtual network. An example of such dependencies is shown in the Figure 2.

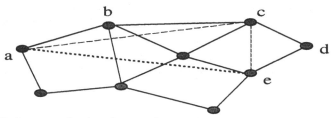

Fig. 2. An example virtual network

Three layers within a multi-layer virtual network are partially represented in the figure above. The link *ae* represents a PVC. This link depends upon links in the layer that supports it, in this case the T1 layer represented by links *ac* and *ce*. These links, in turn, depend upon links in the T3 layer. In the case of link *ac*, its dependencies include links *ab* and *bc*. The link *ce* depends upon the T3 links *cd* and *ce* for its operational definition. An agent-oriented solution to the PVC configuration problem can be found in [16].

6   Agent System Architecture

In the system described here, ant-inspired agents solve problems by moving over the nodes and links in a network and interacting with "chemical messages" deposited in that network. Chemical messages have two attributes, a label and a concentration. These messages are stored within VMCs and are the principal medium of communication used both between swarms and individual swarm agents. Chemical messages are used for communication rather than raw operational measurements from the network in order to provide a clean separation of measurement from reasoning. In this way, fault finding in a heterogeneous network environment is more easily supported. Also, chemical messages drive the migration patterns of agents, the messages intended to lead agents to areas of the network which may require attention. Chemical labels are digitally encoded, having an associated pattern that uses the alphabet {1, 0, #}. This encoding has been inspired by those used in Genetic Algorithms [17] and Classifier Systems [18]. The hash symbol in the alphabet allows for matching of both one and zero and is, therefore, the "don't care" symbol.

Agents in our system can be described by the tuple, $\mathcal{A}=(\mathcal{E}, \mathcal{R}, \mathcal{C}, \mathcal{MDF}, \textit{m})$. This definition is described at length in [19] and will only be briefly described here. Agents can be described using five components:

- emitters ($\mathcal{E}$),
- receptors ($\mathcal{R}$),
- chemistry ($\mathcal{C}$),
- a migration decision function ($\mathcal{MDF}$),
- memory ($\textit{m}$)

An agent's emitters and receptors are the means by which the local chemical message environment is changed and sensed respectively. Both emitters and receptors have rules associated with them in order that the agent may reason with information sensed from the environment and the local state stored in memory. The chemistry associated with an agent defines a set of chemical reactions. These reactions represent the way in which sensed messages can be converted to other messages that can, in turn, be sensed by other agents within the network. The migration decision function is intended to drive mobile agent migration and it is in this function that the foraging ant metaphor, as introduced by Dorigo, is exploited. Migration decision functions have the following forms:

$$p_{ij}^k(t) = F(i,j,k,t) / N_k(i,j,t), \quad R < R^* \tag{2}$$

$$= S(i,j,t)$$

$$N_k(i,j,t) = \Sigma_{j \text{ in } A(i)} F(i,j,k,t) \tag{3}$$

$$F(i,j,k,t) = \Pi_p [T_{ijkp}(t)]^{-\alpha kp} [C(i,j)]^{-\beta} \tag{4}$$

$$F(i,j,k,t) = \max_j \Pi_p [T_{ijkp}(t)]^{-\alpha kp} [C(i,j)]^{-\beta}, \quad j = j^{max} \tag{5}$$

$$= 0$$

where:

$p_{ij}^k(t)$ is the probability that the kth agent at node i will choose to migrate to node j at time t,

α_{kp}, β are control parameters for the kth agent and p^{th} chemicals,

$N_k(i,j,t)$ is a normalization term,

$A(i)$ is the set of available outgoing links for node i,

$C(i,j)$ is the cost of the link between nodes i and j,

$T_{ijkp}(t)$ is the concentration of the pth chemical on the link between nodes i and j for which the kth agent has receptors at time t,

R is a random number drawn from a uniform distribution $(0,1]$,

R^* is a number in the range $(0,1]$,

$S(i,j,t)$ is a function that returns 1 for a single value of j, j^*, and 0 for all others at some time t, where j^* is sampled randomly from a uniform distribution drawn from $A(i)$,

$F(i,j,k,t)$ is the migration function for the kth agent at time t at node i for migration to node j,

j^{max} is the link with the highest value of: $\Pi_p [T_{ijp}(t)]^{-\alpha kp} [C(i,j)]^{-\beta}$.

The intention of the migration decision function is to allow an agent to hill climb in the direction of increasing concentrations of the chemicals that a particular agent can sense, either probabilistically (equation (4) for $F(i,j,k,t)$) or deterministically (equation (5) for $F(i,j,k,t)^2$). However, from time to time, a random migration is allowed, this being the purpose of $S(i,j,t)$. This is necessary, as the network is likely to consist of regions of high concentrations of particular chemical messages connected by regions of low or even zero, concentrations of the same chemicals.

Finally, memory is associated with each agent in order that state can be used in the decision-making processes employed by the agent.

---

[2] $p_{ij}^k(t) = 1$ for $j=j^{max}$ and 0 otherwise.

6.1   Agent Classes

The agent classes defined in the system described here are intended to implement an active diagnosis system [20]. In active diagnosis systems, monitoring and diagnostic activity is undertaken by agents working in a distributed manner in a sensor network and these activities are performed on a timely basis rather than just when a fault is detected. Ishida also describes an immunity-based agent approach to active diagnosis that exploits the metaphor of an immune system for active diagnosis. In some sense, a fault finding system can be thought of as an immune system and agent classes as examples of B-cells and T-cells.

The agent system described here consists of four agent classes. First, condition sensor agents (CSAs) are defined. A CSA is an example of a netlet. The function of a CSA is to measure one or more parameters associated with a given component and determine whether a specific condition is true or false. CSAs interact with VMCs on network components by measuring parameters associated with the network component; e.g. the utilization of links connected to the node or the utilization of node itself. CSAs are adaptive and learn to (a) avoid components where no valid sensory information is available and (b) visit components more frequently that are likely to cause the condition of interest to evaluate to true. While the first situation appears strange at first reading, it must be noted that we are dealing with heterogeneous networks where parameters supported by one vendor may not be supported or provided by another[3]. Therefore, it is likely that CSAs will be vendor specific or apply to a subset of all components in the network at best. Also, it is intended that our CSAs should be self-configuring. Being netlets, they are injected into a mobile code region from a network management workstation and are not directed to visit particular components. It is essential, therefore, that CSAs are capable of learning a map of the network. A CSA's ability to modify the frequency with which it visits a component facilitates variable frequency polling of components. The more the condition for a CSA evaluates to true, the more likely the agent is to visit the component. In this way, CSAs spend more of their processing effort on components with potential performance problems rather than allotting equal time to all components. A CSA may also leave chemical messages on devices that it visits. In this way it is possible for two such agents, one for device type one and the other for device type two, to measure different parameters but generate the same chemical message for use by the fault finding agents. The separation of measurement from reasoning is clearly an advantage here.

It is worth noting that CSAs are capable of interacting with the old manager/agent schema for network management. This can easily be implemented using VMCs. For example, an application that uses a local VMC and implements an SNMP protocol handler can be installed inside the MCD. Thereafter, it can act as an SNMP agent.

Another possibility that has been implemented within the MCE is a handler of an extension protocol. The DPI protocol was chosen for implementation. The DPI

---

[3] A review of the private part of an SNMP MIB for a small number of devices confirms just how diverse devices can be.

protocol was chosen as it is a 'lightweight' protocol and avoids the BER encoding/decoding that is part of SNMP. In this research, a VMC extension registers with an SNMP agent and, acting as an SNMP subagent, provides data in response to SNMP requests. This scenario is shown in Figure 3.

Both of these ideas could also be applied in situations where inter-working with a legacy system is required. It is possible to associate simulated network components with actual devices running legacy agents through properly engineered VMCs. This might be the situation where the actual device does not support a Java environment. It is also helpful within a research environment to be able to link simulated components to the real ones if an idea that has already been tested through a simulation is to be tried on a live network.

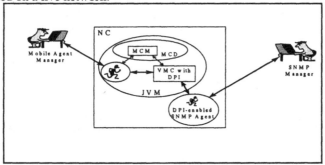

Fig. 3. A sensor agent talks to an SNMP agent

Second, service monitoring (SMA) and service change agents (SCA) are defined. A service monitoring agent is responsible for monitoring characteristics of a set of instances of a service; e.g. the quality of service on one or more PVCs. These agents are static and reside where the service is being provided; e.g. at the source of a PVC. A service monitoring agent detects changes in the characteristics of the monitored service and, if the change is considered significant, a service change agent is sent into the network in order to mark the resources on which the service depends with a chemical message. The concentration associated with the chemical message reflects the change in value of the characteristic of the monitored service. If the change in the measured characteristic for the service is considered beneficial, a negative concentration will be associated with the chemical message; i.e. the chemical will be 'evaporated'. If the change in the measured characteristic for the service is considered detrimental to the service, a positive concentration will be associated with the chemical message; i.e. an existing trail will be reinforced or a new one created. Given that resources will be shared by multiple services, it is easy to see that the resources common to two services will see twice the change in chemical concentration when the SMA detects a significant change. It is this process of chemical interference that allows localization of a fault to be inferred.

Problem identification agents -- other netlets that circulate continuously throughout the network -- use the trail of chemical messages laid down in the network in order to determine the location of faults and to initiate diagnostic activity. These agents form the final class of agents defined. The value of communicating problems to network operators rather than a stream of alarms has long been understood [21, 22, 23]. In this previous work, a static knowledge base system has been developed where the knowledge base is composed of a set of problem classes with communication by messaging between them. A problem class represents a model of one or more potential faults in the network. Instances of problem classes are intended as hypotheses regarding a fault in the network and a winner-take-all algorithm, where the instance explaining the most alarms is considered the most likely problem, is used to discriminate between competing hypotheses.

Mapping a single problem class to a problem agent, and using inter-agent communication for inter-problem message passing, seems a natural progression of this work. Rather than being alarm driven as reported in previous research, problem agents respond to the chemical messages laid down in the network and migrate from component to component based upon the concentrations associated with these chemical messages.

Several problem agents have been implemented. First, a PVC *Quality Of Service* problem agent has been built. This agent hill climbs in the space of the chemical laid down by SCAs. Currently, these agents initiate diagnostic activity on a component when a concentration threshold is reached and this threshold implies that at *least* two SCAs have visited the component. However, reinforcement learning techniques are being investigated as a mechanism for on line learning in order to reduce the number of potential false diagnoses occurring. Diagnostic actions are initiated by interaction with the component through a VMC. When such activity is initiated, the concentration of the 'chronic-failure' chemical, or c-chemical, is increased on the component. A *Chronic Failure* problem agent has been defined in the system that senses the c-chemical for the purpose of identifying components that experience multiple faults in short periods of time. In order that c-chemical concentrations do not increase unchecked, a CSA has been included in the system that periodically visits components and 'evaporates' c-chemical concentrations.

Finally, an *Overload* problem agent has been defined. This agent hill climbs in the space of the concentration of a chemical generated by CSAs that circulate in the network, monitoring component and link utilization parameters. Again threshold driven, it is intended that persistently over-utilized components are identified in order to facilitate re-planning of the network.

7 Conclusions and Future Work

This paper has presented an architecture for a multi-agent system that relies on Swarm Intelligence and, in particular, trail laying behavior in order to locate faults in a communications network. This architecture promotes the idea of a clear separation of sensing and reasoning amongst the classes of agents used and promotes the idea of active diagnosis. A chemically inspired messaging system augmented with an exploitation of the ant foraging metaphor have been proposed in order to drive the mobile agent migration process. The paper has demonstrated how fault location determination can arise as a result of the trail-laying behavior of simple problem agents. An implementation of this architecture has demonstrated that mobile agents can be effectively used to find faults in a network management context. Research continues in the area of adding new problem agents in order to improve the capabilities of the system.

While Java environments have yet to appear on network communication devices, this paper has shown how the architecture presented can be mapped onto a mobile code environment and has briefly described one such environment. The authors are convinced that such mobile code environments will be forthcoming for network components in the *very* near future and management solutions will be based upon them.

While this paper has alluded to a fully distributed network management solution, the authors do not believe this to be entirely desirable. Hence, research described here is currently being extended in order to have a network model -- a core facility required in any comprehensive network management solution -- constructed in real time by the actions of mobile agents. It is intended that, unlike proprietary network model solutions, future research will lead to intelligent network models where behavior as well as state is provided as part of the model and that such behavior is acquired as a consequence of mobile agents visiting a network management workstation. The authors believe that an intelligent network model interacting with deglets and netlets in the network will lead to simpler, scalable network management solutions in the future.

Acknowledgements

We would like to acknowledge the support of the Telecommunications Research Institute of Ontario (TRIO) and the National Science and Engineering Research Council (NSERC) for their financial support of this work.

8 References

Case, J. D., Fedor, M., Schoffstall, M. L. and Davin, C., Simple Network Management Protocol, RFC 1157, May 1990.

Yemini, Y., The OSI Network Management Model, IEEE Communication Magazine, pages 20-29, May 1993.

Yemini, Y., Goldszmidt, G. and Yemini, S., Network Management by Delegation. In The Second International Symposium on Integrated Network Management, Washington, DC, April 1991.

Case, J.D., and Levi, D. B. , SNMP Mid-Level-Manager MIB, Draft, IETF, 1993.

Kotay, K. and Kotz, D., Transportable Agents. In Yannis Labrou and Tim Finin, editors, Proceedings of the CIKM Workshop on Intelligent Information Agents, Third International Conference on Information and Knowledge Management (CIKM 94), Gaithersburg, Maryland, December 1994.

Beckers R., Deneuborg J.L. and Goss S., Trails and U-turns in the Selection of a Path of the Ant Lasius Niger. In J. theor. Biol. Vol. 159, pp. 397-415.

Susilo, G., Bieszczad, A. and Pagurek, B., Infrastructure for Advanced Network Management based on Mobile Code, Proceedings IEEE/IFIP Network Operations and Management Symposium NOMS '98, New Orleans, Luisiana, February 1998.

Bieszczad, A. and Pagurek, B., Network Management Application-Oriented Taxonomy of Mobile Code, to be presented at the IEEE/IFIP Network Operations and Management Symposium NOMS'98, New Orleans, Louisiana, February 1998.

OMG, Mobile Agent Facility Specification, OMG TC cf/xx-x-xx, 2 June 1997.

Beni G., and Wang J., Swarm Intelligence in Cellular Robotic Systems, Proceedings of the NATO Advanced Workshop on Robots and Biological Systems, Il Ciocco, Tuscany, Italy.

Grassé P.P., La reconstruction du nid et les coordinations inter-individuelles chez Bellicoitermes natalenis et Cubitermes sp. La theorie de la stigmergie: Essai d'interpretation des termites constructeurs. In Insect Societies, Vol. 6, pp. 41-83.

Dorigo M., V. Maniezzo and A. Colorni, The Ant System: An Autocatalytic Optimizing Process. Technical Report No. 91-016, Politecnico di Milano, Italy.

White T., Routing and Swarm Intelligence, Technical Report SCE-97-15, Systems and Computer Engineering, Carleton University, September 1997.

Schoonderwoerd R., O. Holland and J. Bruten, Ant-like Agents for Load Balancing in Telecommunications Networks. Proceedings of Agents '97, Marina del Rey, CA, ACM Press pp. 209-216, 1997.

Di Caro G. and Dorigo M., AntNet: A Mobile Agents Approach to Adaptive Routing. Tech. Rep. IRIDIA/97-12, Université Libre de Bruxelles, Belgium, 1997.

Pagurek B., Li Y., Bieszczad A., and Susilo G., Configuration Management In Heterogeneous ATM Environments using Mobile Agents, Proceedings of the Second International Workshop on Intelligent Agents in Telecommunications Applications (IATA '98).

Goldberg, D., Genetic Algorithms in Search, Optimization, and Machine Learning. Reading, MA: Addison-Wesley, 1989.

Holland, J. H., Escaping Brittleness: the Possibilities of General-Purpose Learning Algorithms applied to Parallel Rule-Based Systems. In Machine Learning, an Artificial Intelligence Approach, Volume II, edited by R.S. Michalski, J.G. Carbonell and T.M. Mitchell, Morgan Kaufmann, 1986.

White T., and Pagurek B., Towards Multi-Swarm Problem Solving in Networks, Proceedings of the 3rd International Conference on Multi-Agent Systems (ICMAS '98), July 1998.

Ishida, Y., Active Diagnosis by Immunity-Based Agent Approach, Proceedings of the Seventh International Workshop on Principles of Diagnosis (DX 96), Val-Morin, Canada, pp. 106-114, 1996.

White, T., and Bieszczad, A., The Expert Advisor: An Expert System for Real Time Network Monitoring, European Conference on Artificial Intelligence, Proceedings of the Workshop on Advances in Real Time Expert Systems Technology, August, 1992.

White T., and Ross, N., Fault Diagnosis and Network Entities in a Next Generation Network Management System, in Proceedings of EXPERSYS-96, Paris, France, pp. 517-522.

White T. and Ross N., An Architecture for an Alarm Correlation Engine, Object Technology 97, Oxford, 13-16 April, 1997.

INCA: An Agent-Based Network Control Architecture

Jan Nicklisch, Jürgen Quittek, Andreas Kind, and Shinya Arao

C&C Research Laboratories, Berlin
NEC Europe Ltd.
inca@ccrle.nec.de

Abstract. This paper describes the design and implementation of INCA, an open architecture for the distributed management of multi-service networks and systems applications. The *Intelligent Network Control Architecture* is populated by stationary and mobile intelligent agents. These agents perform monitoring and control of network and systems components, thereby supporting the integrated management of networks and services. The architecture provides transaction capabilities to control transport and mobility of agents, agent prioritization, and multiple agent code transfer schemes. Managed objects used to access resources on network elements and new system functionality can be created, distributed, and replaced dynamically. An example INCA application demonstrates that prioritized agents are necessary to support the timely execution of critical tasks. The design of our agent based network management platform is not bound to a particular programming language or computing environment; the current implementation however, is based on Java and RMI.
Keywords: Mobile Agents, Network Management Platform

1   Introduction

The centralized or hierarchical approaches to network management, being tightly coupled to the client-server paradigm, show well known limitations. Facing the growing complexity and extension of computer and telecommunication networks, it is commonly agreed that there is need to study technologies for distributed network management. The goal is to be flexible, scalable, and open to extensions.

One of the promising approaches is intelligent software agent technology, which supports not only distributed applications but also application mobility. However, despite the research efforts in this area, there is not much acceptance of agent-based techniques in the practical network management world[1]. One of the reasons might be that available agent platforms are general in their design and not customized towards this domain.

As part of the activities at the NEC C&C Research Laboratories in this direction, an open platform for distributed management of multi service networks, INCA (Intelligent Network Control Architecture), has been developed. The most important

---

[1] Exceptions are SNMP and CMIP agents. They are widely used but incorporate a very restricted application of agent technology.

difference between INCA and other agent platforms, among several customizations, is reliability.

This paper gives a description of the techniques applied to the design of our architecture. INCA is based on software agents, which can migrate over the network to perform delegated and/or mobile functions in an intelligent manner. There is also support for open signaling, allowing flexible service creation as well as extensibility and modular replacement of elements, services and functions.

The remainder of this paper is organized as follows: after discussing related work in the following section, we motivate the use of software agents, intelligence, and mobility in management of networks and services in Section 3. The INCA architecture is introduced in Section 4, and the features which make it a reliable platform, well-suited for network management and telecommunications applications, are discussed. Measurements demonstrating the usefulness of agent prioritization are presented in Section 5.

2 Related Work

The general advantages of decentralized and agent-based approaches to network management and telecommunications have often been addressed, for example see [1, 2, 3, 4, 5].

Currently, several mobile agent environments are available, mostly due to the wide acceptance of the Java environment with its inherent support for code mobility; see [6] for an overview. Four of the major environments for agent-oriented Java programming are: *Aglets* [7], *Concordia* [8], *Odyssey*[2] (a Java-based subset of a platform formerly known as *Telescript* [9]), and *Voyager*[3]. However, all these and many others are general purpose environments and not customized for network management, where reliability and efficiency are important issues. We argue that for high reliability transaction capabilities for interactions of distributed entities are required. Based on such safe interactions, a fault-tolerant control of the itinerary of a mobile agent is possible. For efficiency, agent prioritization and support of different agent code transfer schemes is desired.

Concrete network management applications focusing on agent intelligence have been described by da Rocha and Westphall [10] and Somers [11], both using stationary agents.

Sahai et al. have presented an agent environment customized for network and system management, called Astrolog [12]. They employ mobile agents to support the mobility of the network operator. However, the core network management system is based on a hierarchy of stationary agents.

Surprisingly, even in the network management domain there is no flexible environment for mobile intelligent agents offering the means for reliability and efficiency mentioned above.

---

[2] Odyssey home page: http://www.genmagic.com/agents/odyssey.html
[3] Voyager home page: http://www.objectspace.com/voyager/index.html

3   Network Management with Agents, Intelligence and Mobility

To ensure effective and efficient function of large networks as well as provision of services, and furthermore to address the requirements of network providers, service providers, retailers and users, there is a strong need for a coherent framework supporting automated and more intelligent end-to-end management solutions.

Network management today is typically based on a client-server model with SNMP [13] and CMIP [14] as the de facto standards for monitoring and for managing a network of computers and devices. Managed objects, like hosts, routers, bridges, switches, printers etc., provide read/write access to a set of variables through a management process. A management station can then communicate with the management processes in a client-server relationship.

Although simple, this centralized approach to network management has some severe drawbacks with regard to scalability, flexibility and performance. By dividing management functions into mobile, autonomous and intelligent computing entities (i.e. software agents), many of the problems with network and service management can be addressed:

- Scalability is increased, as management is not performed solely by the management station but delegated to distributed management agents.
- Repetitive tasks can be avoided if software agents learn from experience.
- Fault tolerance can be increased through the autonomy and learning capability of management agents.
- Better performance is achieved by moving the management functionality closer to the actual network element, thus reducing network traffic.
- The low-level details of different devices can be hidden behind the agent interface.
- Legacy systems can be integrated by using an agent for inter-operation.

Typical scenarios like service provision through different administrative domains, including accounting and charging, can be handled by a multi-agent based approach to distributed network and service management in a more scalable and coherent way than with a centralized approach to network management.

This is particularly true for the creation of new services in the telecommunications world. With the advent of high-speed networks based on ATM switching, there is demand for better support when creating and managing new services. Proprietary, low-level interfaces to network devices currently prevent a rapid development of telecommunication applications (see [15]). Control and management of scalable multi service networks is difficult since the switching software is usually tightly coupled with the individual switching devices.

TINA [16], xbind [17] and DCAN [18, 15] focus on providing more coherent and flexible access to multi service networks. However, control and management of switching devices based on mobile software agents has not been investigated so far.

4 The Architecture

INCA can be described as a mobile agent platform customized for network management and control. This section gives an overview of the architecture and discusses the major design decisions. In particular, we will explain the concepts of agent priorities, transaction capabilities, migration control, and multiple agent code transfer schemes, which in their combination distinguish INCA from other available agent environments.

4.1 General Architecture

An instance of the INCA platform consists of a set of stations and of services offered locally by a station or globally by the platform.

4.1.1 Local Services

A station is local to a network element and provides the following services:

- *Concurrent execution of multiple agents.* Multiple agents—each with its own thread of control—can be executed concurrently. The agents are scheduled according to their priority attribute by an INCA specific scheduling scheme, which for example takes into account the current system load. No agent of lower priority will be executed if there is an agent with higher priority in a 'runnable' state present at a station.
- *Loading of agent code.* Agent code can be transferred between stations. Three schemes for code transfer are supported: push, pull, and migrate. Pushed code is sent from an agent code repository to a dedicated station, pulled code is loaded by the station from a repository, and in migrating code is sent from one (non-repository) station to another. All stations support the same code format, i.e. the same code can be executed at any stations. In order to improve performance, caching of agent code is used.
- *Transfer of agent state.* The current state of an agent can be transferred from one station to another. In combination with agent code transfer, this service enables agent migration.
- *Transaction capabilities.* For peer-to-peer communication between stations a reliable and fault-tolerant layer is used offering transaction capabilities.
- *Access to local resources.* Access to local resources at a station is provided by *managed objects.* Managed objects represent network elements (or parts of these) that are to be managed. Their interface specifies which resource can be accessed and how. Usually, a station contains at least one instance of a managed object giving access to the resources of the local network element. Additional managed objects can be created dynamically on demand of an agent.
- *Support for location and status monitoring of agents.* Location and status of an agent migrating from station to station may be monitored by a central instance

called *locator*. For monitored agents the station send messages to the locator on arrival and departure or termination of the agent.

4.1.2 Global Services

Besides services provided by each single station there are services offered by the entire platform, or by a set of dedicated stations:

- *Control.* For each INCA platform there is at least one *control station*. This station executes a stationary control agent offering a control service for launching and monitoring agents. Usually, the location of this station is the console of the network manager.
- *Repository.* Dedicated stations offer a repository service. This service provides access to the code off all agents that can run on the platform. The repository can be accessed via the control service to push agent code to a station or by any station to pull agent code.
- *Locator.* A unique station runs a stationary locator agent offering the location tracking service. The locator agent receives messages about monitored agents from the stations. Based on this information the locator agent answers requests from the control agent about location and status of other agents. Furthermore, the locator agent supervises predefined itineraries (see below).
- *Itinerary control.* The control service allows to link a mobile agent with a predefined itinerary. The itinerary becomes part of the agent state, but is not accessed by the agent itself. It is controlled by a station whenever a mobile agent is to be migrated, in order to determine the station to send it to. When an agent with a predefined itinerary is launched by the control agent, a copy of the agents itinerary is sent to the locator agent which compares the itinerary with location messages.
- *Migration control.* By combining itinerary control with persistence of agent state fault tolerance can be increased for the migration of agents.
- *Inter-agent communication.* Communication between agents is offered by a common interface for inter-agent message invocation. This functionality includes a naming service from which agents can get references to other agents.

4.2 Priorities

Concurrent execution of multiple agents does not only increase the functionality of an agent platform, it also introduces the problem of scheduling agents. We observed that a fair and equal scheduling of all agents in a system does not match the requirements of network management applications. The groups of agents listed below illustrate these requirements, for example.

- *Network Control Agents* support the network manager e.g. in configuration and fault management of the network. The tasks they pursue are demand driven, perhaps triggered by network faults, and often have to be carried out as fast as possible.

- *Service Management Agent* support the service provider, e.g. by a service subscription agent. They use the network environment together with the platform services to install and establish end user services, or reconfigure them.
- *Network Maintenance Agents* perform repeatedly occurring management tasks, e.g. the gathering of data from selected network elements or fault analysis. Their tasks are carried out repeatedly, only rarely requiring adjustments by the network management instance.
- *Network Monitoring Agents* perform routine tasks such as the filtering of raw data collected from network elements. Often, such agents are stationary, since their task is to monitor a concrete element of the network, or a link.

Usually, network control agents should be executed fast, because the network manager is waiting for the execution. In contrast to this, monitoring agents typically run for a long time and should be interrupted, when one of the other kinds of agents arrives to be executed.

We use agent priorities to satisfy these requirements. An arriving agent of higher priority interrupts an agent of lower priority and no agent of lower priority is executed while an agent of higher priority is present and not blocked.

By these means the problem above, can be solved by giving network control agents the highest priority, service management agents and network maintenance agents a medium priority, and monitoring agents the lowest priority.

An application described in Section 5 demonstrates the usefulness of the priority-based approach.

4.3 Transaction Capabilities

Most interactions in INCA are peer-to-peer communications between stations. In the platform we have a clear separation of different layers of communication facilities. INCA is designed to run on top of an object-oriented distributed middleware providing naming service and transparent access of objects regardless of their physical location.

Between the middleware and the application, INCA contains another layer called *transaction capabilities* providing reliable, fault-tolerant interactions between two stations. The design of this layer is based on the SS7 (Common Channel Signaling System No.7) transaction capabilities [19]. SS7 transaction capabilities use distributed transaction monitors to ensure the reliable exchange of messages between two communicating peers.

In the INCA platform we used this technique to support asynchronous exchange of messages as well as synchronous exchange. On the application level these two are sufficient to express all interactions of two stations conveniently. Retransmission attempts of messages, time-outs, health checks, and further means providing reliability interactions.

4.4 Migration Control

When starting an agent with a predefined itinerary, a copy is stored at the locator. A failure at the station currently hosting the mobile agent does not necessarily lead to a breakdown of the agents task, since an up-to-date copy of the agent's itinerary is kept, and a new agent can be created in order to visit the remaining stations of the itinerary. Of course, this fault-tolerant procedure should only be applied to agents, which have been designed according to the procedure. In contrast to this concept, mobile agents might not make use of such an itinerary at all, but instead determine the hosts to be visited dynamically, at runtime. It depends on the particular application, which of both methods should be employed. We observe that applications of mobile agents, which require high reliability, as it is the case for many network management tasks, can make good use of the itinerary concept. To understand why that is the case, consider the following example.

Suppose that a mobile maintenance agent is used to perform a routine task at all the network elements, or a fixed subset of it. An example could be an agent, which examines the network elements and checks the current memory usage and the disk space used. In case of an alarming situation at the network element, it will send a message to the network manager, otherwise it proceeds along its itinerary. In such an example there is actually no state involved in the agent migration, apart from the information contained in the itinerary. Now consider a breakdown at one of the intermediate stations. Without any observing instance the work done by the mobile agent, i.e. the confirmation of a certain system state at the stations visited, would be lost. The platform would have to instantiate another agent and start from scratch, because it cannot be aware of how far the agent could proceed its task. And in case certain adjustments were made by the agent at the network elements, these would perhaps have to be overwritten.

Since the system employs the locator, however, it can recover from such simple error cases and let another instance of the maintenance agent proceed the task from the station following the one which experienced the breakdown. Excessive system and network load can be avoided, because this second instance does not need to repeat its predecessor's work.

Apparently, the idea of keeping a copy of an agent's itinerary is not the full story, but only the most simplified explanation of a more general recovery scheme. When the agent state actually matters, as it is the case for more complicated or more intelligent agents, the knowledge of the itinerary alone is insufficient. Instead, the whole agent state should be preserved at any source station of an agent migration. Then, after the destination station experiences a fault, the locator instance times out in waiting for the message that confirms the agent's arrival. It can therefore take further actions and restore the agent from the state previously cached at the source station.

4.5 Multiple Agent Code Transfer Schemes

INCA uses mobile code for all agents. For stationary agents the agent code has to be transferred to a single station only, whereas mobile agents require the code on all visited stations. When migrating, the state of a mobile agent has to be transferred from the station the agent is leaving to the station the agent is going to visit next. But since the code of the agent is independent of the current agent instance and its state, it can be transferred in different ways, and it can be cached at stations. INCA supports three schemes for code distribution: pull, push, and migrate.

Push type agent code distribution. This scheme is known from Internet services like *PointCast*[4] or *Marimba's Castanet*[5]. While these services push user data to a list of subscribers, the push scheme in INCA pushes agent code to a list of stations. Usually, the push scheme is in conjunction with predefined itineraries. When the agent is launched, the code is pushed to all stations of the itinerary. So, when the agent instance arrives at a station, the corresponding code is already there and it can start execution immediately. The push scheme is realized by sending a message to the repository station, which pushes the code to the stations.

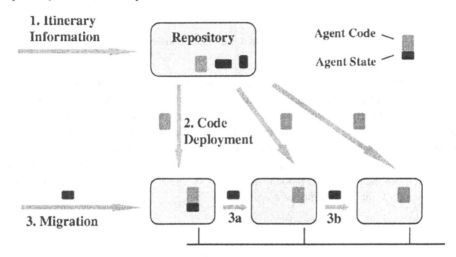

Fig. 1. Push type agent code distribution

Pull type agent code distribution. The pull scheme is probably the most widespread one, since it is used by web browsers to download Java applets and scripts. When this scheme is used, a station downloads the corresponding agent code from a repository station after an agent instance has arrived. This scheme can be applied to mobile agents with flexible itineraries. These agents may decide dynamically which station to visit next. The drawback of this scheme is a decrease of performance. When an agent arrives at a station, its execution has to be delayed until the agent code has been downloaded. Agent code caching can avoid this delay for subsequent arrivals of agents of the same type. An alternative would be pushing the code of the agent to all stations, but this solution reduces scalability, produces unnecessary load on the network, and might increase memory requirements of the stations.

---

[4] PointCast Network: http://www.pointcast.com
[5] Marimba's Castanet: http://www.marimba.com/new

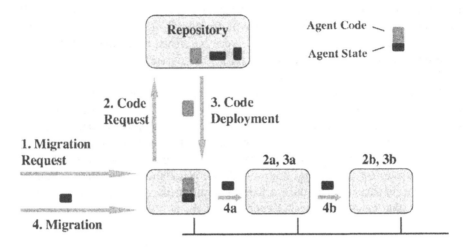

Fig. 2. Pull type agent code distribution

Migrate type agent code distribution. This scheme is the most obvious one to use for mobile agents. Code is moved together with the agent instance from station to station. Usually, the performance is lower than the push scheme performance, but higher than the pull scheme performance. Compared to the push scheme the migration time is longer, because agent instance and code have to be transferred. Compared to the pull scheme, communication with an repository station is only required when launching an agent. After launch the agent is more flexible, because of being independent of the availability of a repository.

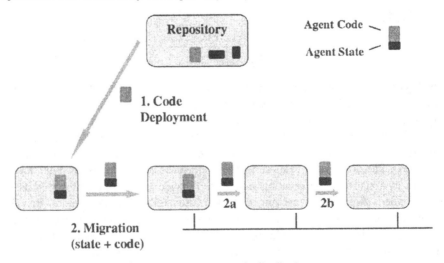

Fig. 3. Migrate type agent code distribution

Since there are advantages as well as shortcomings in all of these three distribution schemes, it depends on the application, on the type of task the mobile agent has to work on, which scheme is preferable. The availability of more than one scheme also gives more flexibility for cache management. All stations provide caching of agent code, in order to improve agent performance and to reduce communication load on the network. But since the cache size is limited, cached code has to be dropped from time to time. Now, if push type agent code has been loaded, this code may be dropped by the cache manager—if required—even before the agent which is going to use it has arrived. In this case, the agent code is loaded corresponding to the pull scheme when the agent arrives. There are good reasons to consider dropping pushed agent code from the cache before the agent arrived. The agent might be delayed because of congestion in the network or because of its low priority, or it might have terminated irregularly. In the latter case it will never arrive at some stations and it would be wrong to keep the code in the cache infinitely.

4.6   Implementation

The current implementation of INCA is based on the Java language. Java already supports code mobility and dynamic class loading. However, we replaced the Java class loader in order to be able to load code from the repository station and to support multiple code mobility schemes. As communication infrastructure we use Java RMI, but CORBA is also supported.

Agents are implemented as Java threads with agent priorities translated into Java thread priorities. This mapping follows the INCA system policy imposed on various agent types (see Section 4.2), and takes into account the load level of the current station (e.g. the memory usage). As such it is more powerful than Java's priority-based scheduling. Agent instance transfer is based on Java object serialization. The state of an agent is represented by an object, which can be sent from station to station in its serialized representation.

A set of managed objects is provided as a library. Network elements like workstations or servers which can host an INCA station are supported by managed objects with direct access to the network element. Network elements without capabilities to run a Java virtual machine, like routers and ATM switches are supported by proxy managed objects accessing these elements via the local area network. However, our experiences with INCA showed that for many applications co-development of agents and managed objects is desirable.

5   Impact of Prioritization

In order to demonstrate the usefulness of agent prioritization for network management applications, we chose a simple scenario. It consists of a set of connected LANs. At each LAN there is one network element hosting an INCA station (see Figure 4). Each INCA station provides a set of managed objects allowing agents to access the network elements in the LAN. Furthermore, we consider a management station launching agents for monitoring and control. Monitoring agents are stationary and monitor dedicated network elements of a LAN. Thus, there typically is more than one monitoring agent at a station. Different to these continuously running monitoring

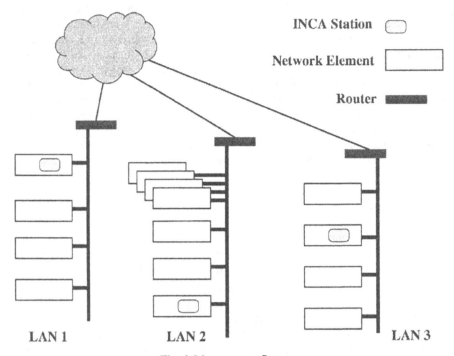

Fig. 4. Measurement Setup

agents, there are control agents which are launched interactively by a network operator to perform a usually short task. For these agents instantaneous task completion is desired.

Figure 5a) shows measurements with three INCA stations, each of them hosting the same variable number of monitoring agents. These agents all have the same priority. Then a control agent is launched to visit each of these stations once to perform his task. If this agent has the same priority as the monitoring agents (dashed line) the time required for its completion increases linear to the number of threads per station. If otherwise he has a higher priority than the monitoring agents (solid line) the time required for its execution is almost independent from the number of monitoring agents and the reaction to the operator is always as fast as without any monitoring agents.

Figure 5b) shows a similar measurement. Here the number of threads per station is a constant of two for all stations, but the number of stations is varied. Again, the advantage of high priorities for control agents can be observed. The measurements were conducted on Windows NT 4.0 workstations with Pentium 166 processors, connected by 10BaseT Ethernet.

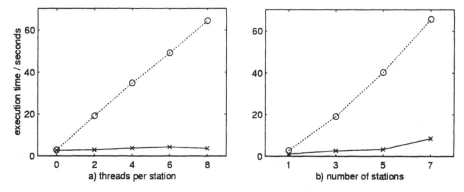

Fig. 5. Execution time of a control agent

6 Conclusion

INCA, the Intelligent Network Control and Management Architecture, is an infrastructure to support applications in the integrated network and service management and telecommunication areas. INCA is especially targeted at distributed management applications based on intelligent mobile agents. This technology overcomes several of the inherent difficulties of centralized, client-server based management systems.

The architecture provides prioritized agents, transaction capabilities, migration control, and multiple agent code transfer schemes to support the creation of highly reliable and efficient applications. The impact of agent prioritization has been demonstrated for an example network management application.

7 Future Work

So far, INCA agents are implemented in a straightforward manner, with the algorithms coded directly into them. Although this approach is sufficient for smaller applications and quick prototyping, we are going to extend the INCA platform by libraries for agent intelligence. Hence, our main work is currently focused on choosing a proper goal representation and problem solving formalism.

Also we are aware of the importance of security in agent systems, in particular in network management. Various security mechanisms have already been proposed, but it is still subject to research which of them is suited best for mobile agents in network management.

Acknowledgments

This work was carried out as part of our efforts towards maintainable high speed networks for multimedia communications, at the NEC C&C Research Laboratories in Berlin. We are grateful to S. Iwasaki and N. Elshiewy for valuable feedback and comments, and we would like to thank S. Robidou for implementing INCA applications and performing the measurements used in this paper.

8 References

[1] G. Goldszmidt and Y. Yemini. 1995. Distributed Management by Delegation, from *15th International Conference on Distributed Computing Systems*. IEEE Computer Society.

[2] K. Meyer and M. Erlinger and J. Betser and C. Sunshine and G. Goldszmidt and Y. Yemini. 1995. Decentralizing Control and Intelligence in Network Management, from *Integrated Network Management IV.*, ed. A. S. Sethi and Y. Raynaud and F. Faure-Vincent Chapman Hall, pp. 4-16.

[3] T. Magedanz and K. Rothermel and S. Krause. 1996. Intelligent Agents: An Emerging Technology for Next Generation Telecommunications?, from *IEEE INFOCOM 1996*.

[4] P. Charlton and Y. Chen and E. Mamdani and O. Olsson and J. Pitt and F. Somers and A. Wearn. 1997 (February 5-8). An Open Agent Architecture for Integrating Multimedia Services, from *Proceedings of the 1st International Conference on Autonomous Agents.*, ed. W. Lewis Johnson and Barbara Hayes-Roth ACM Press, New York, pp. 522-523.

[5] M. Baldi and S. Gai and G. P. Picco. 1997. Exploiting code mobility in decentralized and flexible network management, from *Proc. First Int. Workshop on Mobile Agents.*, ed. K. Rothermel and R. Popescu-Zeletin Springer-Verlag, Berlin, pp. 13-26.

[6] J. Kiniry and D. Zimmerman. 1997. A Hands-on Look at Java Mobile Agents. *IEEE Internet Computing*, 1(4), pp. 21-30.

[7] D. B. Lange and M. Oshima and G. Karjoth and K. Kosaka. 1997 (mar). Aglets: Programming Mobile Agents in Java, from *1st Int'l Conf. on Worldwide Computing and Its Applications '97 (WWCA97).*, ed. T. Masuda, Y. Masunaga and M. Tsukamoto SV.

[8] D. Wong et al. 1997. Concordia: An Infrastructure for Collaborating Mobile Agents, from *Proc. First Int. Workshop on Mobile Agents.*, ed. K. Rothermel and R. Popescu-Zeletin Springer-Verlag, Berlin, pp.86-97.

[9] J. E. White. 1994. *Telescript Technology: The Foundation for the Electronic Marketplace*. Technical report. General Magic, Inc.. 2465 Latham Street, Mountain View, CA 94040.

[10] M. A. da Rocha and C. Becker Westphal. 1997. Proactive Management of Computer Networks using Artificial Intelligence Agents and Techniques, from *The Fifth IFIP/IEEE Int'l Symposium on Integrated Network Management*. Chapman Hall, pp. 610-621.

[11] F. Somers. 1996 (August). *HYBRID: Intelligent Agents for Distributed ATM Network Management*.

[12] A. Sahai and S. Billiart and Ch. Morin. 1997 (February). Astrolog: A Distributed and Dynamic Environment for Network and System Management, from *Proceedings of the 1st European Information Infrastructure User Conference*.

[13] J. Case and M. Fedor and M. Schoffstall and J. Davin. 1990 (may). A Simple Network Management Protocol (SNMP); RFC-1157. *Internet Request for Comments* (1157).

[14] W. Stallings. 1993. *SNMP, SNMPv2 and CMIP: The Practical Guide to Network-Management Standards.* Addison-Wesley. Reading.

[15] K. van der Merve and I. M. Leslie. 1997. Switchlets and Dynamic Virtual ATM Networks, *IM '97.*

[16] L. A. De la Fuente and J. Pavon and N. Singer. 1994. Application of TINA-C Architecture to Management Services. *Lecture Notes in Computer Science,* 851.

[17] A. Lazar. 1997. Programming Telecommunication Networks. *IEEE Network,* pp. 8-18.

[18] S. Rooney. 1997. The Hollowman an innovative ATM control architecture, *IM '97.*

[19] ITU-T. 1993. Signalling System No.7 - Functional Description of Transaction Capabilities. ITU-T Recommndation, March 1993.

Agent Metaphors for Analysing Telematic Services

Guy Davies and Love Ekenberg

Department of Information Technology (ITE), Mid-Sweden University
S-851 70 Sundsvall, SWEDEN
{guy, love}@ite.mh.se

Abstract. Using as little mathematical detail as possible, we present the use of transformations between process related languages and dynamic entity-relationship models. The work incorporates our earlier research in schema integration and the integration of multi-agent architecture designs to handle problematic cases of global inconsistency in distributed information systems. The verification methods in first order logic (FOL) presented here are generally applicable to the analysis of conflicts in process based systems. Agents are a useful metaphor for handling asynchronous processes in contexts such as telecommunications. Although the application here is one of telephonic services, the general framework when applied to agent architectures is equally suitable for extended services such as decision making agents as the basis for intelligent home services. Telia Research AB (previously Swedish Telecom) commissioned this work, the implementation of which was used to detect interferences to Telia's public telematic services.

1   Motivation

The agent metaphor has turned out to be most useful for describing complex software artefacts [Shoham 93a]. Using this metaphor, we may consider telematic services as processes with a potential for mental and social properties such as beliefs, knowledge, capabilities, plans, goals, desires, intentions, obligations, commitments, etc. In order to co-ordinate their activities, each agent has an (event-driven) agenda describing what it is to do in the future.

The main purpose of this work is to describe how telematic services represented in a process-driven language can be translated into a logic based formalism so as to prepare a unifying view of a general class of conflict detection. This is demonstrated by a simple type of conflict detection for interactions between common telematic services. Translating specifications of agents into FOL has the added advantage that by introducing 2nd order analyses, it becomes possible to analyse a broader spectrum of conflicts, i.e. using FOL extended with transaction mechanisms provides tools for systematically classifying conflicts, including: individual inconsistency; protocols for multi-agent behaviour; incongruencies when events are incompatible; whether certain combinations of initial states are incongruent; whether event paths are incompatible, etc. cf. [Ekenberg 96a, Ekenberg 98a]. This kind of unified view is made possible through translations of the kind suggested in this paper, even when the processes are described in an event-driven representation. In the same way, the framework proposed herein can be used to handle more elaborated types of personalised telematic services and analyse how these can be integrated with a working system.

The main part of this paper presents the use of transformations from representations of agents as processes to their corresponding dynamic entity-relationship models. This is exemplified by showing how interference can be detected between various kinds of telematic services as specified in SDL. We argue that FOL when used in conjunction with conceptual modelling, provides a sound basis on which specifications written in a process based language can be transformed, merged, and verified for the purpose of detecting interferences.

There are two problem areas which our work comes to grips with. The first concerns difficulties in implementing telematic service specifications of disparate nature. The second concerns difficulties in verifying agents behaviour when specification analysis leads to state space explosion. The problem of disparate specifications arises from the fact that they are often developed by different people or work units and may therefore be expressed in different terms or structures. This naturally increases the difficulty of co-ordinating communication between the various models and the data they use. Decentralised, architectures of autonomous agents, that have been developed independently therefore create the need to integrate design models and co-ordinate information. Another side to this problem is that different portions of telecom systems may have been developed at different times. Furthermore new sub-systems must continuously be integrated with older systems in a way that will not lead to conflicts or other undesired effects. An important part of conceptual design in this context is the integration of various conceptual schemata. The term *schema integration* will be used to refer to this process, which is more precisely defined in [Batini 86]. The telematic services upon which this work is based were specified in the process based language SDL. To enable the various methods for schema integration to be used and to allow analysis of the integrated schema, specifications had to be transformed into a conceptual model. Only then was it possible to carry out certain kinds of analyses on the merged specifications.

The second problem addressed in this work lies in the fact that when, as is common, agents are specified in the form of asynchronous processes, analysing their behaviour becomes difficult. For instance, techniques for analysing SDL specifications such as those described in [Holzmann 92], allow perhaps 95% of the state space to be analysed. However, with no probability distribution available over the state space, states belonging to the unexplored 5% may have a high probability of occurring. An approach to reducing the state space explosion problem is to found in [Heinrich 90] which simplifies extended finite state machines to one minimal finite state machine. However, this does not allow schema merging. Other translation approaches which do not however address the problems of verification and schema merging are to be found in [Fröberg 93] which transforms SDL to the declarative language Erland and [Cheng 90] which provides a transformation frame work. Our approach in this paper differs from these by taken its standpoint that the transformation should serve to preserve the conceptual integrity of the original specification as well as to provide a sound basis for both merging and verification.

For these reasons we chose to model in first order logic (FOL). This is a language that is more fully understood that lends itself more readily to in-depth analyses and at the same time being attractive from both a conceptual and computational viewpoint.

By allowing a process based specification to be transformed into a conceptual model expressed in FOL it then becomes possible to analyse the entire state space. This approach is therefore particularly suited to applications where safety is crucial. The method takes advantage of the fact that analysis of a conceptual schema over a determined finite domain results in satisfiability problems in first order propositional logic. The importance of transforming process based specifications into a conceptual model is two-fold: it allows not only the creation of a consistent set of specifications using techniques in schema comparison, unification and integration; but it also allows exhaustive verification to be carried out. It is the bridging between these two forms of specification that constitutes the main substance of this work.

Work on conflicts between agents has frequently been based on communication between agents and most of the literature is focussed on resolving conflicts. This frequently requires a precompiled frame work of co-ordinating plans, as worked on by [Shoham 93b, Moses 90] or expensive run-time protocols for co-ordinating negotiations between agents [Rosenschein 94, Sycara 88, Kuwabara 89, Kraus 91] and altering belief systems as proported in [Rao 95]. However, tackling the difficulties in detection and identification of conflicts has received notably less attention. This is in itself a complex task.

Rather than further investigate established approaches such as those above, we identify conflicts between agents through the use of schema integration techniques. This is achieved by modelling service agents in FOL and eliminating possible conflicts by integrating them into more comprehensive services. Our approach is constructive in that it results in new integrated services from previously disparate ones; and diagnostic in that it identifies conflicting states as well as routes by which they can be reached. For details see [Ekenberg 96e].

2   Preliminaries

An essential observation in this context, of the difference between process based and logic based models of a system, is that in the process based model the static properties of the system are contained, and thereby concealed, within the internal states of its component agents, whereas in a system based on FOL the system's properties are immediately apparent by examination of the formulae. Likewise the dynamic aspects of process based specifications display a courser granularity for 'atomic' behaviours whereas in FOL every change to the state of the system can be analysed for consistency.

2.1   Conceptual Modelling

In the definitions below, we assume an underlying *language* L of first order formulae. We also assume a basic knowledge of conceptual modelling. The unfamiliar reader is referred to [Boman 97].

A conceptual schema usually contains a static and a dynamic part. The static schema expresses the model's static characteristics, the state space of the model using graphs with additional textual formulae.

By a *diagram* for a set R of formulae in a language L, we mean a Herbrand model of S, extended by the negation of the ground atoms in L that are not in the Herbrand model. Thus, a diagram for L is a Herbrand model extended with classical negation.[1]

A *schema* S is a structure <R, ER> consisting of a *static part* R and a *dynamic part* ER. R is a finite set of closed first order formulae in a language L. ER is a set of *event rules*. Event rules describe possible transitions between different states of a schema and will be described below. *L(R)* is the *restriction of L to R*, i.e., L(R) is the set {p | p ∈ L, but p does not contain any predicate symbol, that is not in a formula in R}. The elements in R are called *static rules* in L(R).

The dynamic part of the schema is a collection of transition rules, each of which specifies how a state description in the schema is changed. Intuitively, an event gives rise to a transition from one diagram to another one, and is the result of an event rule together with an event message in the system or its immediate environment. Event rules consume signals and change the state of process entities. Below, *L* denotes a language and *B* denotes the alphabet underlying the language L.

2.1.1   Definition:
An *event message* in L is a vector **e** of constants in the alphabet B.

2.1.2   Definition:
An *event rule* in L is a structure <**z**, P(**z**, **w**), C(**z**, **w**)>. P(**z**, **w**) and C(**z**, **w**) are first order formulas in L, and **z** and **w** are vectors of variables in B.[2] P(**z**, **w**) denotes the precondition of the event rule, and C(**z**, **w**) the post condition.

2.1.3   Definition:
An *event* for a **schema** <R, ER> is a tuple (σ, ρ), where σ and ρ are diagrams for R, and

(i)   ρ = σ, or
(ii)   there is a rule <**z**, P(**z**, **w**), C(**z**, **w**)> in ER, and an event message **e**, such that σ is a diagram for P(**e**, **w**) and ρ is a diagram for C(**e**, **w**). In this case we will also say that (σ, ρ) *results* from the event rule and the event message.

Note that an event rule is non-deterministic: if a diagram σ satisfies the precondition of an event rule then any diagram ρ satisfying the event rule's post condition will give rise to an event (σ, ρ) resulting from the event rule.[3]

---

[1]For our purposes, this is no loss of generality by the well-known result that a closed formula is satisfiable iff its Herbrand expansion is satisfiable. For a discussion of this expansion theorem and its history, see, e.g. [Dreben 79].

[2]The notation A(x, y) means that x and y are free in A(x, y).

[3] In this way the approach to the dynamics of a schema presented in this paper differs from the traditional transactional approach of the database area, where an event deterministically specifies a minor modification to a state [Abiteboul 88].

3 The Transformation

This section provides an intuition of how a subset of SDL can be transformed directly into a conceptual schema. The reader unfamiliar with SDL is referred to [Belina 91]. The work described in this paper is based on a subset of SDL-88 as defined in [CCITT 88]. The language's basic structural properties are all present in this version. In spite of the later attempt in SDL-92 to add object oriented features (requiring more than double the number of non-terminals), SDL-92's formal properties remain the same as in '88, and are sufficiently expressive for our purposes.[4]

3.1 Pre-processing

The strengths of SDL that its modularity provide are of no interest for the purposes of translation into a conceptual schema or what follows. Our basic assumption is that process types in SDL correspond to entity types in a conceptual schema. The only additional information required for modelling a nested specification is extra signal routes. Blocks can be eliminated and the processes they contain made to belong to one and the same level, as long as the following constraints on communication in an SDL system, as illustrated in Figure 1, are preserved in translation.

1. Channels and routes have direction
2. A particular signal type can be allowed to travel both ways along a particular channel or route.
3. A particular signal type can travel along completely different channels or routes
4. A particular signal type can travel along a particular route or channel for part of its journey but along different routes for the initial or final part of its journey.

Several strategies for decomposing this structural form are possible but the simplest have the draw back of producing either redundant paths that duplicate the same route for different signals or spurious paths such as r2_c1_r3, which although it is a possible path does not actually conduct any signal.

The renaming procedure suggested here preserves the channel route combinations symbolically in the names of the routes. Processes are ensured unique names by prefixing them with the name of the block in which they were previously contained. Channels and signal routes are treated in a similar way. This allows both a static manual check of the flattened structure against the original SDL specification as well as dynamic check in a simulation. The latter compares permitted routes for any given signal type with those routes that actually occur as signal instances in a queue thereby ensuring that no signals go astray.

---

[4] The reader should be aware that the presentation in this section is simplified for the purpose of emphasising the principles for the transformation. A more rigorous treatment and exposition is to found in [Davies 96].

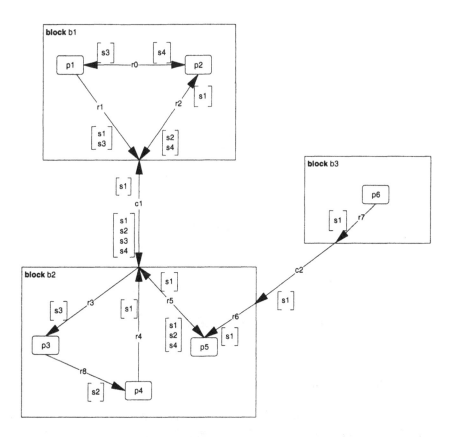

Fig. 1. A simple route and channel structure

The intuition behind the flattening algorithm is illustrated by Figures 1 and 2. A series of channels and routes between source and destination process types are transformed into one unified route between source and target. Paths from source to target that branch are converted into as many unique routes as there are branchings. Bi-directional channels and routes are treated as two uni-directional routes.

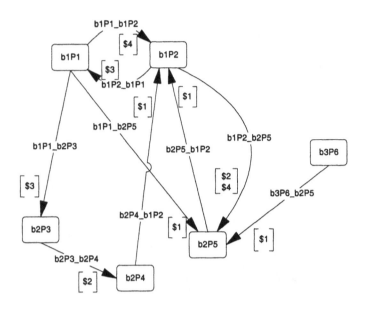

Fig. 2. The flattened structure of Fig. 1

3.2 Static Schema Transformation

The mapping algorithm that converts the flattened SDL specification into a conceptual model is a syntax directed translation [Aho 72, sec.3.1] that traverses the syntax tree using a top down, depth first strategy implemented using a tail recursive procedure.[5]

After translation the complete static schema has an appearance similar to Figure 3. The entity types that correspond to process types are prefixed with createdProcess. The super type createdProcess serves to make the model more compact. Instances of this type may exist in exactly one of the two states running or terminated. Entity types processType, route and signalType allow the dynamic checks mentioned above

Provided the number of process that may be created during an execution is limited, the transformation to the conceptual model can be done by representing existing processes and non-existent processes in two different states in the entity type. The creation of a process is achieved by changing the state of the process instance in the conceptual schema from terminated to existent. Later when the process dies, the instance returns to its former state of terminated.

Signal types are treated similarly. The relation precedes which is used to derive the predicate first which identifies which signals lie at the head their respective queues. The predicate succeeds is necessary to prevent the same signal form being used twice by the same rule.

---

[5] The formal details of this transformation are given in [Davies 96].

The entity types queuingSignal_S1 have a relation, the range of which is the entity type createdProcess_P2. This indicates that the type of value contained by queuingSignal_S1 is that of a process instance of createdProcess_P2. In general, variables of type PId are represented by a relation to the entity type createdProcess.

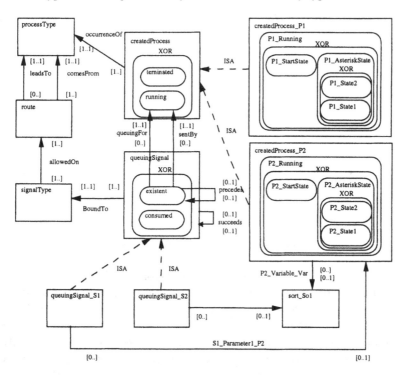

Fig. 3. The conceptual model of the process based language SDL[6]

Sorts in SDL can also be modelled naturally by entity types given the restriction that the number of elements of a particular sort must be finite. Finally variables, parameters and the properties of signals, all become relations that vary over the instances of their respective entity types.

3.3   Dynamic Schema Transformation

The dynamic part of the conceptual schema models the sending of signals to an addressed process using a rule in the conceptual schema that adds the signal to the queue belonging to that process

When a signal is consumed from a queue, all elements in the queue are moved forward and a transition occurs (in general). This can be handled with the aid of an information processor that handles signal consumption and state transition

---

[6] For the details of the particular modelling notation used in figure 3, see [Widebäck 92] and [NP/FW/001].

represented as rules. The equivalent applies to operations occurring between two state transitions.

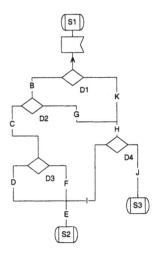

Fig. 4. A simplified decision structure in SDL

In order to illustrate the ideas behind the translation algorithm let us look at the decision structure of an SDL transition. Although the decision structure shown in Figure 4 is simplified, it covers all aspects necessary to illustrate the principles of the translation. Decisions are represented by rhombi and letters on the decision paths are intended to represent 'atomic' or transactional sequences of <action>s between divergences or convergences of such paths. Divergence and convergence amount to the same thing as far as the specification of an event rule is concerned: They both extend the transition path and divide it into 'atomic' sections. The <action>s in these sections include variable assignments and decision alternatives, signal sending, and process creation.

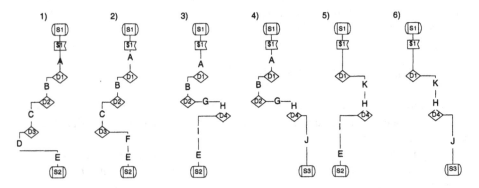

Fig. 5. Decomposed decision graph of Fig. 4 showing all unique paths emanating from state S1

In the translation, each unique path through an SDL process graph that emanates from one state symbol and finishes in another will correspond to one event rule in a conceptual schema. Figure 5 shows the decomposition of Figure 4 into its unique paths and reveals something of the intuition behind the translation of the dynamic part of SDL into a conceptual schema.

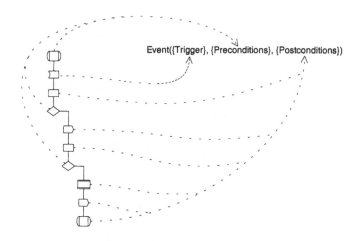

Event({Trigger}, {Preconditions}, {Postconditions})

Fig. 6. Depositing SDL <action>s into an event rule

Figure 6 shows very schematically how different elements in the decomposed decision graph are transformed into different elements in an event rule.

The way that the dashed arrows join one another is intended to indicate the algorithmic order in which components from the transition path are added to their respective sets. Likewise the way that arrows unite before they arrive at their respective sets of conditions is intended to indicate that the order of components has no meaning in the context of their new representation.

<[],

(queuingSignal_drop_hook_a(QSx) **AND** telephone1_connected(Px) **AND** first(QSx, Px) **AND** consumed1(CQSx) **AND** queuingSignal_on_hook(CQSx) **AND** nw1_Running(Destination)),

(consumed(QSx) **AND** sentBy(CQSx, Px) **AND** existent(CQSx) **AND** queuingFor(CQSx, Destination) **AND** telephone1_idle(Px))>

Fig. 7. A simple transition expressed in SDL and as an event rule.

Figure 7 shows an example of how a path is translated. The process is in the state Connected until the signal Drop_hook_A is first in the queue to the process and is consumed. When this happens the signal On_hook is sent to the process NW1 upon which the process changes state to Idle. The event rule is internal_event, which is generated by the information processor. When the rule is triggered the processor checks whether there is an instance of drop_hook_a first in the process' queue. It also checks whether the process is in the state Telephone1_connected. (This represents the state Connected and in the conditional part it is prefixed with the process' name.) following this the processor checks whether there is a signal instance of type on_hook that is not already queuing for a process. This is achieved with the predicate consumed1(CQSx) and queuingSignal_on_hook(CQSx). In the condition part is also tested whether there is an instance of Destination for the signal instance On_hook. The destination captured in the condition is then used in the conclusion part of the rule. When the conclusion part of the rule is executed the predicate sentBy(CQSx, Px) becomes a fact in the information base, which thereby contains information about which process sent the signal. The predicate existent(CQSx) provides the information that the signal instance CQSx is not available to be sent elsewhere. The predicate queuingFor(CQSx, Destination) represents the fact that the signal instance is queuing for the process Destination. The conclusion part also contains the predicate Telephone1_idle(Px) which represents the fact that the process has entered the new state Idle. It also contains the predicate consumed(QSx) which indicates that the signal QSx is consumed and is again available for sending.

4   Detecting Service Interference - An Example

For the system to trigger an event rule there must be a signal at the front of a process's queue. This is stated as a precondition. An event rule's post conditions dictate the changes made after it has fired. These include specifying how sent signals should be placed in the queue of their target process. In many cases the receiver of a signal needs to be able to find the identity of the sending process. This is achieved by leaving the triggering signal in the queue while a transition is being processed so that its sender can still be identified. In addition to the conditions mentioned are all the conditions necessary to fire and tidy up after an implicit transition. The remainder of this section demonstrates detection of service interference using an implementation of the work described above. Interference detection is based on analyses of the merged specifications of two telephone services.[7]

An information processor handles the state transitions in a conceptual schema. It also carries out static and dynamic consistency analyses for conflictlessness as well as allowing queries to be answered. Because the techniques used to perform the analyses of the conceptual model are based on theorem proving, the information processor was based on SICSTUS PROLOG and uses an implementation of Stålmarck's algorithm [Stålmarck 95a]. The processor begins by checking the initial instantiation is in keeping with the conceptual schema. During initial execution it displays the following:

> Parsing file bc.pr...
> Preprocessing SDL specification...
> Translating SDL specification...
> Creating instantiation...

---

[7] The details of merging models can be found in [Ekenberg 96d]. Also to be found in this document are the SDL specifications that underlie the transformation and the subsequent execution explained below.

By enriching the transformed schemata with more invariants, further conditions can be imposed. The invariants are written in the form of typed formulae in first order logic. First order formulae are translated into propositional logic. Typing is carried out in order to minimise the length of the resulting propositional logic formulae. A transformation of this kind is possible since the schema contains a finite number of instances of each type of entity. A universally quantified expression can therefore be expressed as a conjunction and an existentially quantified expression as a disjunction.

The following invariant for example expresses that if one subscriber's phone is ringing then another subscriber's phone must be calling. The characters $procD ensuing the quantifier mean that the variables x and y may only be instantiated with instances of processes. Typing of this kind is defined during transformation.

> [ALL,$procD,(telephone1_ringing(X) | telephone2_ringing(X) | telephone3_ringing(X) |
> telephone4_ringing(X)) ->
> [SOME,$procD,(telephone1_calling(Y) | telephone2_calling(Y) | telephone3_calling(Y) |
> telephone4_calling(Y))]]

The following series of events causes the system to reach a forbidden state that violates the condition stated in the invariant above. This is equivalent to a service interference.

Subscriber b rings and speaks to subscriber c, during which time subscriber a rings up subscriber b, upon which subscriber b receives a signal that a call is waiting. Subscriber b however decides not to take the call from subscriber a. Instead b disconnects the waiting signal from a, without interrupting his conversation with subscriber c. This is expressed as follows:

> send_signal(lift_hook_b).
> send_signal(digits_b(i3)).
> send_signal(lift_hook_c).
> send_signal(lift_hook_a).
> send_signal(digits_a(i2)).
> send_signal(cw_ignore_b).

Event rules are fired once their preconditions are satisfied. Following the execution of each event rule the updated facts are displayed and the information processor checks whether the new state satisfies conceptual schema. The following text shows an excerpt from the screen's display.

> Checking invariants:
> .... Queued signals (2): [network_ask_pid_2]
> ** Queuing network_ask_pid_2 for nw1
>   Signals being first: [network_ask_pid_2-nw1]
>   Queued signals (1): [network_ask_pid_2]
> rules([[2017]])
> rule(2017)
> consumed(network_ask_pid_2)
> existent(network_answ_pid_2)
> nw1_init2(nw1)
> queuingFor(network_answ_pid_2,conndb)

sentBy(network_answ_pid_2,nw1)
~consumed(network_answ_pid_2)
~existent(network_ask_pid_2)
~nw1_initial(nw1)

When the conditions of an invariant are not satisfied by the new state as dictated by the most recently executed event rule, the following type of message is displayed:

Checking invariants:
....ERROR: False invariant
all(dprocD,_173317,telephone1_ringing(_173317)#telephone2_ringing(_173317)#telepho
ne3_ringing(_173317)#telephone4_ringing(_173317) =>
some(dprocD,_173293,telephone1_calling(_173293)#telephone2_calling(_173293)#telep
hone3_calling(_173293)#telephone4_calling(_173293)))

If the violated invariant was added during schema merging, it can be concluded that the merged schema gives rise to interference.

5   Concluding Remarks

This paper proposes a working process for the analysis of SDL-specifications. Most detail has been devoted to the translation into conceptual schemata. The translation is crucial in this context as it allows for the use of theorem proving techniques for analysing transaction systems, in accordance to their static and dynamic properties. The approach we have devised is to model the basic functional components of a chosen subset of SDL in an entity-relationship model. The translation of any SDL specification within this subset of the language then becomes a question of extending the base model with subtypes together with their relations, according to the SDL specification's use of basic SDL components. The resulting model is expressed in a format of first order logic.

The paper also demonstrates how interferences in telecom services, originally written in a process based language, were detected using the translated specifications. This was achieved using simulation techniques. An alternative approach to detecting disturbances in such services might be to examine whether it is at all possible to arrive at a forbidden state by way of an event rule, given a starting state that is in keeping with the static schema. In this way, the entire state space can be explored. The way in which this kind of analysis can be carried out on conceptual models is described in more detail in [Stålmarck 95b]. The use of translations for analysing process related specifications by theorem provers for first order logic can also define more elaborated services. For instance by integrating the authors' earlier research in the integration of multi-agent architecture designs to handle problematic cases of global inconsistency in systems, cf. [Ekenberg 96b, 97; Danielson 98, Ekenberg 98].

Acknowledgements

This work was supported by the Graduate School of Teleinformatics.

6 References

[Abiteboul 88]   Abiteboul, S. & Vianu, V. "Equivalence and Optimization of Relational Transactions", Journal of ACM, 35(1), pp. 130-145, 1988.

[Aho 72]   Aho, A. V. & Ullman, J. D. *The Theory of Parsing, Translation, and Compiling*, Prentice-Hall 1972.

[Batini 86]   Batini, C., Lenzerini, M. & Navathe, S. B. "A Comparative Analysis of Methodologies for Database Schema Integration", ACM Computing Surveys, vol. 18, no. 4, pp. 323-364, 1986.

[Belina 91]   Belina, F., Hogrefe, D. & Amardeo, S. *SDL with Applications from Protocol Specification*, Carl Hanser Verlag and Prentice Hall International (UK), 1991.

[Biskup 86]   Biskup, J. & Convent, B. "A Formal View Integration Method", in International Conference on the Management of Data, Washington, ACM, 1986.

[Boman 97]   Boman, M., Bubenko, J., Johannesson, P. *Conceptual Modelling*. Prentice Hall. 1997

[CCITT 88]   CCITT Recommendation Z.100: Specification and Description Language SDL (Blue Book, Volume X.1-X.5, 1988, ITU General Secretariat-Sales Section, Places des Nations, CH-1211 Geneva 20).

[Dayal 84]   Dayal, U. & Hwang, H.-Y. "View Definition and Generalisation for Database Integration in a Multidatabase System", IEEE Transactions on Software Engineering, vol. SE-10, no. 6, pp. 628-644, 1984.

[Danielson 98]   M. Danielson and Ekenberg L., "A Framework for Analysing Decisions under Risk," to appear in European Journal of Operations Research, 1998.

[Davies 96]   Davies, G, R. "A Pilot Study of a Translation from SDL-88 to a Language for Conceptual Modelling", technical report, Telia Research AB. Stockholm 1996.

[Dreben 79]   Dreben, B. & Goldfarb, W. D. "The Decision Problem: Solvable Classes of Quantification Formulas", Reading, Mass, Addison-Wesley, 1979.

[Ekenberg 96a]   Ekenberg L. and Johannesson, P. "A Formal Basis for Dynamic Schema Integration," Proceedings of 15th International Conference on Conceptual Modelling ER'96, pp.211-226, LNCS, 1996.

[Ekenberg 96b]   Ekenberg L., M. Danielson, and M. Boman, "From Local Assessments to Global Rationality," *International Journal of Intelligent and Co-operative Information Systems*, vol. 5, Nos. 2 & 3, pp.315-331, 1996.

[Ekenberg 96c]   SDeLphi: Detection of service interference through the use of formal methods, Ed. Ekenberg, L., technical report NP-K-LE-OO5, Telia Research AB, Logikkonsult NP AB, 1996.

[Ekenberg 96e]   SDeLphi: Detection of service interference through the use of formal methods, Ed. Ekenberg, L., technical report NP-K-LE-OO5, Telia Research AB, Logikkonsult NP AB, 1996.

[Ekenberg 97]   Ekenberg L., M. Danielson, and M. Boman, "Imposing Security Constraints on Agent-Based Decision Support," Decision Support Systems International Journal, vol.20, No.1, pp.3–15, 1997.

[Ekenberg 98] Ekenberg L., "Goal-Conflicts between Decision Making Agents," to appear in Journal of Logic and Computation, 1998.

[Ekenberg 98a] Ekenberg L. and Johannesson, P. "Detecting Temporal Agent Conflicts" submitted

[Heinrich 90] Heinrich, R., Pospiech, F. "Control Flow Analysis – A New Approach for Avoiding the State Space Explosion Problem", Journal: Informatik, Informationen Reporte No.10, pp 115-123, 1990.

[Holzmann 92] Holzmann, G.J. "Practical Methods for the Formal Validation of SDL Specifications", Computer Communications., Vol. 15 No 2, pp. 129-134, 1992.

[Johannesson 91] Johannesson, P. "A Logic Based Approach to Schema Integration", in 10th International Conference on Entity-Relationship Approach, Ed. T. Teorey, San Francisco, North-Holland, 1991.

[Johannesson 93] Johannesson, P. "A Logical Basis for Schema Integration", in Third International Workshop on Research Issues in Data Engineering - Interoperability in Multidatabase Systems, Ed. H. Schek, Vienna, IEEE Press, 1993.

[Kraus 91] Kraus, S., Ephrati, E., Lehmann, D. "Negotiation in a non-cooperative environment. Journal of Experimental and Theoretical Artificial Intelligence, 3(4):255-282, 1991

[Larson 89] Larson, J. A., Navathe, S. & ElMasri., R. "A Theory of Attribute Equivalence in Databases with Applications to Schema Integration", IEEE Transactions on Software Engineering, vol. 15, no. 4, pp. 449-463, 1989.

[Kuwabara 89] Kuwabara K., Lesser, V.R, "Extended Protocol for Multistage Negotiation" Proceedings of the Ninth Workshop on Distributed Artificial Intelligence, pp 129-161, Rosario, Washington, 1989.

[NP/FW/001] Stålmarck, G., Widebäck, F. "Definition av Delphi" Logikkonsult NP AB, 1991

[Rao 95] Rao, A. S, Georgeff M. P, "BDI Agents: From Theory to Practice" Proceedings: First International Conference on Multi-Agent Systems, pp. 313-319, 1995.

[Rosenschein 94] Rosenschein, J.S, Zlotkin, G, "Rules of Encounter" , The MIT Press, 1994

[Moses 90] Moses, Y, Tennenholtz, M. Artificial social systems part1: Basic principles. Technical Report CS90-12, Weizmann Institute, 1990.

[Shoman 93a] Shoham, Y. "Agent-Oriented Programming", Artificial Intelligence, vol. 60, pp. 51-92, 1993.

[Shoham 93b] Shoham, Y, Tennenholtz, M. On social laws for artificial agent societies: Off-line design. Artificial Intelligence, 1993.

[Stålmarck 95b] Stålmarck, G., Ekenberg, L., Berg, S., Olsson, I. & Nordberg, F. Pilot study of formal analysis of service interaction, validation and verification, Telia Research AB, 1996.

[Sycara 88] Sycara, K.P. "Resolving Goal Conflicts via Negotiations" Proceedings of the Seventh National conference on artificial Intelligence, pp 245-250, St. Paul Minnesota, 1988.

[Widebäck 92] Widebäck, F. "The Graphical Notation of Delphi", Logikkonsult NP AB, 1992.

Multi-agent Testbed and an Agent Launch Tool for Diverse Seamless Personal Information Networking Applications

Suhayya Abu-Hakima[1], Roger Impey[1], Ramiro Liscano[1], and Amir Zeid[2]

[1]Institute for Information Technology, National Research Council of Canada,
Building M-50, Montreal Road, Ottawa, Canada K1A 0R6
web: www.nrc.ca/iit/SPIN_public
suhayya@ai.iit.nrc.ca,
[2] School of Computer Science, Carleton University, Ottawa Canada
zeid@scs.carleton.ca

Abstract. This paper describes the design and implementation of a unique cooperative agents testbed and tool for launching cooperative agents in an environment that addresses diverse applications for the difficult problem of seamless personal information networking (SPIN). The real-world SPIN testbed is aimed at two difficult applications, namely seamless messaging and intelligent network management. Both applications are agent-driven and share agent behaviours and the messaging agents rely on the network management device diagnostic agents for input. The paper introduces both problem areas in a common testbed. The agent launch tool is described in detail. User-centric seamless messaging assumes heterogeneous communication environments intended to support today's nomadic users. The prototype is introduced for the management of messages across distributed information networks. The aim of it is to intercept, filter, interpret, and deliver multi- modal messages be they voice, fax, video and/or e-mail messages. A user's Personal Communication Agent is charged with delivering messages to the recipient regardless of their target messaging device be it a telephone, a pager, a desktop, a wireless laptop or a wireless phone. Personal Communication Agents classify and act on incoming messages based on their content. A Secretary Agent routes and tailors urgent messages appropriately to the Device Manager Agent which delivers the message to a device that the user may be roaming or active on. What makes the Seamless Messaging Application unique is its approach to treating a message in a universal manner, its ability to mediate between different messaging environments and devices, and its ability to try to track and find the user.

1 Introduction

Today's intensively distributed workspaces require users to communicate and compute across heterogeneous networks and applications. The paper describes the design of an agent launch tool for a cooperative agents testbed that aims at addressing applications for the difficult problem of seamless personal information networking. The SPIN testbed is aimed at two difficult applications, Seamless Messaging and Intelligent Network Management. The paper describes the multi-agent framework intended for both problem areas. It also gives details on the first generation seamless messaging prototype.

Users send and receive voice mail, electronic mail (e-mail) and fax over a variety of wired and wireless networks. The paper addresses *Seamless Messaging (SM)* in heterogeneous environments and the design of cooperative agents to address the problem [1]. SM allows users to work in distributed personal workspaces and have messages created and delivered how, when, where, and if they wish them to be. Seamless messaging is defined here as the ability of the user to send and receive messages in a manner transparent to the modality (be it voice mail, email or fax mail), the networks (be it wired or wireless voice or data) and the devices (be it a cellular phone, a pager, a desktop computer, etc.). To achieve such transparency, networks have to be used invisibly. Thus, users have to be located intelligently (through electronic calendars or active devices) and messages have to be tailored for delivery to diverse devices on demand (e.g. messages may have to be converted from long multimedia desktop messages to short voice ones).

Seamless Personal Information Networking (SPIN) is the vision of personal information networking technologies that allow users to seamlessly interchange information in today's distributed workspaces how, when, where, and if they want to. The assumption is that every user has a unique and distributed workspace. In this workspace, a user can have heteregeneous devices (telephones, computers), running on heteregeneous networks (wired, wireless, voice, data, multimedia) and all integrating a variety of heteregeneous applications (voice mail, email, fax, word processors, web browsers, electronic calendars, etc.).

Put simply, Seamless Personal Information Networking is based on the concept that today's user workspace is no longer bound by four walls. Rather, today's workspace is a virtual one with a multitude of applications. Furthermore, people work with networks of desktop and mobile computers, fixed and mobile telephones, fax machines, and pagers, with both hardware and software constantly evolving. SPIN is a simple concept that is quite complex to achieve. To make SPIN a reality, a real-world cooperative multi-agent system (MAS) testbed with an agent launch environment has been put in place to prototype personalised networking applications.

2   Agents in Seamless Personal Information Networking

Many reasons come to mind why agents are an ideal evolution of the traditional distributed computing paradigm for seamless networking. Agents are accepted as software entities that can act autonomously or with some guidance on the user or software systemís behalf. Agents are active computational entities that are persistent, can perceive, reason and act in their environments and can communicate with other agents [2]. Agents are ideal for applications that require some form of distributed intelligent cooperation.

Networks can be seen as a natural domain for the application of distributed artificial intelligence, and more particularly, agent-based computing technology [3]. In particular, Weihmayer and Velthuijsen suggest a number of reasons for this, including their inherent distribution (e.g. along spatial, functional, and temporal lines), the proliferation of heterogeneous devices and services associated with them (this is particularly true of multi-vendor mixed computing-communications networks), the growing need for privacy, the sustained demands for high performance, and the increasing desire for ìintelligenceî in the network [4].

Modelling messaging in organization services as a collection of coordinated agents results in a number of benefits. For example, a degree of virtual homogeneity is brought to otherwise heterogeneous networks of computer-telephony messaging services and devices (such as voice mail, e- mail or fax mail); relatedly, a more open network architecture facilitating more rapid and effective deployment of ìplug and playî messaging services is made possible. All the same, the agent metaphor does not, in and of itself, directly resolve any of the technical issues related to system inter-operability such as sharing remote resources, guaranteeing a particular quality of service or resolving the network feature interaction problem. Rather, as Laufmann points out in [5] ìthe metaphor provides a model of coordination that addresses real-world issues of the computing and communications marketplaces, and in so doing leverages the deployment of new technical solutions as they become availableî.

Users require both heavyweight or complex reasoning processes in the form of personal assistants that can manage their computing and communication needs as well as lightweight or simple processes that can act as proxies on their behalf in the network. As a user roams from place to place, they require distributed support to access their information as the need arises. Active information processes embodied as agents can provide such support. These agents, like those supported by the Carnot projectís Distributed Communicating Agents (DCA) tool [6], are essentially high-performance problem solvers which can be located anywhere within and among networked enterprise resources. These agents are intended to communicate and cooperate with each other, and with human agents. Through the use of models of other agents and resources within the enterprise, SPIN seamless messaging and network management agents, like DCA agents, will be able to cooperate to provide integrated and coherent management of information in heterogeneous computing-communications environments [2].

The convergence of networks and the need for personal digital assistants with embedded agents that interface to all the information accessible through a web of networks (Internet, World Wide Web, etc.) has also been cited as a key reason to develop cooperative multi-agent systems [7]. SM agents are being developed to enable users to share valuable messaging services such as voice and e-mail as if in a single seamless network across a variety of devices that include cellular phones, PDAs, wired telephones and wireless laptop computers.

Several systems have been developed that mirrors in its goals what SPIN is aiming for. These include the Multi-Agent Network Architecture (MANA) described in [8;9], KAoS (Knowledgeable Agent-oriented System) developed   [10], and General Magic's Telescript and Magic Links mobile agents [11]. The agents described in MANA are coarse-grained agents that lack the flexibility to adapt and are a predecessor to those developed for SPIN. KAoS focuses on monolithic reasoning rather than tailoring behaviours based on the context of the their world. General Magic's agents assumed a closed world that did not interoperate with much of the user's workspace (the telephone being the most obvious example). SPIN's agents on the other hand focus on light weight processing, situated personalised functionality, and a focus on user-centric computing and communications.

3   Living Lab: SPINís Real-World Multi-Agent System Testbed

For the SPIN vision to be achieved, a real-world networking environment must be used as the underlying environment for the agent testbed. At the Institute for Information Technology (IIT) we have set up the heteregeneous environment known as the IIT Living Lab. The idea behind the Living Lab is to allow agents to be created, to monitor themselves, to co-exist, to spin-off proxies and to be killed or expire when they have fulfilled their duties. Both the Seamless Messaging and the Intelligent Network Management Application Agents share the testbed. The intent is for the Personal Communication Agents (PCAs) of the seamless messaging application to make use of the Diagnostic Agents (DAs) of the network management application. For example, if a PCA needs to forward information to a user device that is not responding it could ask the device DA or its parent DA ëwhy is the active user device not responding?í.

Figure 1 illustrates the IIT Living Lab. It is centered around a typical enterprise LAN. The Institute LAN has daily operational and experimental traffic for an organisation of over 100 users with hundreds of interconnected devices over seven subnetworks. Initially, we are launching the agents into the SPIN subnetwork which has over 30 desktop devices which are also accessible from home by dialing in. The

SPIN subnetwork also has network devices such as printers, routers, etc. connected to it. The PCAs will typically reside on the user desktops.

In the case of a user with a number of desktops (the average at SPIN is three: two at the office and one at home), the user PCA can reside on a single device but monitor activity on the other devices. This will be possible through the DAs which will reside on every networked device and work as proxy agents of SNMP agents (SNMP is the Simple Network Management Protocol from the Internet Engineering Task Force). SNMP agents are very simple and function to place respective device information in a Management Information Base (MIB). The MIB is very much like a database that can be browsed by network management tools for simple device monitoring information.

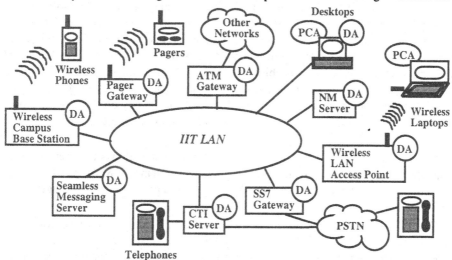

Fig. 1. The IIT Living Lab.

Both the Seamless Messaging application and Intelligent Network Application will have agents resident on a respective server. The idea behind the server is to house any agents that would not be device resident. It is also intended to be the place for the persistent agents to be stored if user devices are shut down (as is typically done with todayís desktop PCs) or if user devices experience problems. Our strategy will be to maintain a backup of the PCA for the user with a scalable time delay (a day-old backup would be the default but if a user wants one on demand it will be provided). DAs will be backed up in a similar manner. Here we take the lessons from the real-world in computing and communications to plan ahead to avert catastrophe.

The IIT LAN has a wireless access point connected to it with its own Diagnostic Agent proxy to its SNMP agent which monitors its operation and access. The wireless access point allows a user to walk around campus with a wireless laptop. Again, a user PCA may be resident on a wireless laptop or may send a PCA proxy to a laptop with a reduced set of behaviours.

Also connected to the LAN is a Computer Telephony Integration platform (CTI) which allows the user to receive telephone calls with any associated telephony information (incoming caller id, call forwarding info., etc.) to the desktop. The LAN is thus connected to the public telephone system so that it may allow the PCA access into voice mail environments that are normally telephone- driven. The CTI platform will also have a diagnostic agent proxied to its SNMP agent.

Furthermore, a wireless campus base station allows a user to roam about with a wireless phone which the PCA can also route calls to. A paging gateway is also accessible through the LAN for the PCA to page the user as instructed. Both the base station and the pager gateway will have diagnostic agents attached to them to monitor their health. Proxy PCAs with simplified behaviours may be dispatched to a cellular phone or pager as long as the device can host it. For the devices to host a PCA proxy, we assume some processing capabilities on board. This will soon be facilitated to some extent as pager and wireless phone manufacturers deliver JAVA-enabled devices with JAVA virtual machines on board. In this manner, the PCA proxy implemented in JAVA can be hosted. At the moment JAVA-enabled pagers and wireless phones are not part of our testbed and we simply rely on device-compatible gateways to deliver text or voice information to the pager or cellular phone.

Finally, an Asynchronous Transfer Mode (ATM) switch allows the user access to high speed multimedia applications such as video on demand. Most ATM switches are now SNMP-enabled and thus our DA can work as a proxy to its SNMP agent.

The IIT Living Lab makes an ideal testbed into which we have launched our seamless networking applications. The Living Labís heteregeneous networks are essentially transparent to the user and are simply used by the personal communication agent to route user information as needed. The networks are also being managed by a set of diagnostic agents which share some behaviours with the personal communication agents. Furthermore, the PCAs rely on some of the monitoring infor- mation from the DAs to fulfil their obligations. Thus, in a *single* testbed, SPIN is ensuring the re- usability of agent behaviours and the interoperability of two agent- based applications. This is a key aspect to our work which we believe is a goal for many researchers but is fairly rare if non existent in other multi-agent system testbeds.

4 SPINís Agent Launch Environment

4.1 What is SPINís Agent Launch Environment?

The agent launch environment is the heart of SPINís multi-agent system testbed. Through this web-enabled environment different types of agents can be launched into the system to perform certain tasks. The ultimate goal of the launch environment is to provide a plug-and-play environment for our SPIN applications. By plug-and-play we mean, adding, installing and removing network elements and services can be done with minimal human effort using mobile and stationary agents, and without affecting the overall performance of the system. Each user will have his own customizable Personal Communication Agent (PCA) that can be launched using this environment. The PCA acts on behalf of its owner and initially sets the user's calendar, equipment, classes of messages and rules to act upon receiving messages. Also through this environment users are expected to shop for their requirements of devices, services, and utilities using a set of mobile agents. This is similar to the proposed multi-agent agent launch environment proposed by [12] based on JAE (Java Agent Environment). JAE is a mobile agent environment built on top of Java and is currently under development at Aachen University of Technology. However, our focus is both static and mobile agent launch while JAE focuses only on mobile agents.

Using object-oriented design/analysis methods seemed like a good starting point to model agents; however, due to differences between objects and agents a different modelling process was needed. We are following a simple process [13] to find and model agents in the agent launch environment. The process is a combination of some of the well-known object-oriented analysis/design methods and it adds some ideas to adapt the nature of agents.

The agent launch environment, figure 2, has a user friendly interface with different views according to the authorization level of users. General options include shopping, adding and removing utilities, devices and services. These tasks will be achieved by the interaction and communication of the underlying agents. In addition, we provide a monitoring facility to keep track of users (through their PCAs), agents and devices.

Fig. 2. The agent launch environment.

4.2 Types of users who interact with the agent launch environment

This section introduces types of users who interact with the system from the services point of view. Roles are defined according to TINA's specifications [16] with slight modifications. The Telecommunications Information Networking Architecture (TINA) Consortium is an international consortium aiming at defining and validating an open architecture for telecommunications systems for the broadband, multi-media, and information era. TINA, which is based on distributed-objects concepts, is mainly composed of four architectures: service, network, management and computing. TINA integrates all the control and management functions into a unified, logical software architecture. When TINA project is finished in its current form, the communications industry will have at its disposal an open software architecture with a complete set of specifications for building and managing services on a global scale.

Three types of users can interact with SPINís agent launch environment: a user who is normally a consumer or subscriber; an enterprise which is formed by a group of users and a network manager who administers the services and registrations. The three types of users supported will interact with the service as follows:

User Type
A User can interact with the service as follows:
User A establishes a Service Session including User B.
User modifies User service profile.
User shops for personal device
User adds, removes personal network device.

Enterprise Type
An Enterprise is a group of users who use the service in a similar way; however, each user will have freedom to personalize his service profile. An Enterprise can interact with the service as follows:
Enterprise adds, removes User as service user.
Enterprise cancels subscription for a service.
Enterprise modifies User service profile.
Enterprise modifies Global service profile for all subscribed Users.
Enterprise monitors Service Users.

Network Manager Type
A Network Manager can interact with the service as follows:
Network Manager adds, removes Service Provider to network.
Network Manager shops for network resource, service, utilities.
Network Manager adds, removes network resource.

4.3 SPIN main agents in the launch environment

In this section we will introduce some of the main agents in SPIN. The list is by no means complete, we are introducing the main agents that we will be using in the example scenario (section 6.4). Conversion, communication, PCA, services and shopping agents are integral to the seamless messaging application while network elements and service agents are integral of the network management application.

4.3.1 PCA

The Personal Communication Agent (PCA) is a scripted agent whose role is to manage all the user's messaging needs. Its role is to accept a message and have it classified then acted on by a classifier and action behaviors.

4.3.2 Conversion Agents

Conversion agents are provided by service providers. They are mainly used to transform messages from one format to another. Figure 3 shows the hierarchy of conversion agents. Examples include:

Text-Speech Agent
This agent converts a text string to an utterance through a speech synthesis facility.

Fax-Text Agent
This agent takes an encoded fax message as input and decodes it to an image of 1ís and 0ís. The text on the image is then recognized through optical character recognition.

Video-Audio Agent
This agent streams the audio from an encoded video. The audio is represented as a WAVE file so that it may be used in standard messaging environments and be digitally encoded.

180

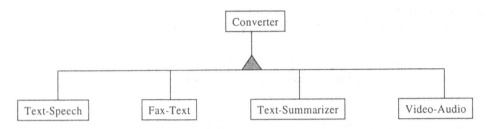

Fig. 3. Converter hierarchy.

4.3.3 Connection Agents

Connection agents are provided by service providers. They provide connections that the PCA may require to arrive at the user in a certain format. Figure 4 shows the hierarchy of connection agents. Examples include:

WWW Connector
This agent provides on demand access to the WWW. It may use either Netscape or Microsoft Explorer as the web connection based on the environment variables it is provided to link to.

Pager Connector
This agent provides access to a 2-way paging connection on demand. As a secondary function, the pager connector can also send a page command with a set of digits to an external paging service.

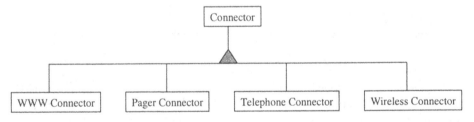

Fig. 4. Connector hierarchy.

4.3.4 Network Elements Agents

Network elements agents are provided by network vendors. They are used to manage network resources since they represent network elements. The ultimate goal of this type of agents is to add robustness and ease of use to the system. Elements can be added and removed from the network without affecting the system as a whole. Each element has an agent that interacts with network managers. Figure 5 displays the hierarchy of device agents.

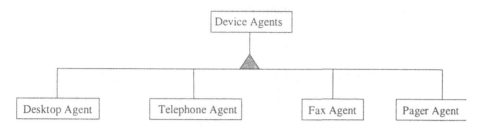

Fig. 5. Device agent hierarchy.

4.3.5 Service Agents

Service agents are provided by service providers. They are mainly used to manage services. The main tasks of this type of agents are to install, uninstall and update services with minimal human interaction.

4.3.6 Shopping Agents

Shopping agents are controlled by users and network managers. They are mainly used to search and locate users requirements and then give advice according to userís criteria. This set of agents are mobile and will wander on the Internet to fetch userís requirements. Figure 6 shows the hierarchy of shopping agents

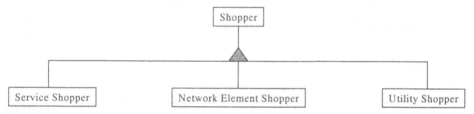

Fig. 6. Shopper hierarchy.

4.4 Example Scenario

In the following scenario we will show how a user can add a new network element using the agent launch environment. Before adding a component (a cellular phone for example), the user may want to shop for the best available phone using the shopper agents. After finalizing the choice, the user is now ready to install his component, so the component-installer agent will start interacting with other agents (device manager, converters and connectors) to achieve the task. The detailed scenario is shown below:

4.4.1 Scenario for adding a component

1-User requests adding a personal component

2-User shops for the component (if desired) (Figure 7)

2.1 User enters his preferences

Name/description(e.g. Cellular Phone)Price   (e.g. $100)

Vendors   (e.g. Nokia, Ericson)

Already existing utilities   (e.g. Microcell, Bell Mobility)

2.2 Shopper agent moves on to available vendors

2.3 Shopper agent comes back with a list of available devices

2.4 User finalizes decision

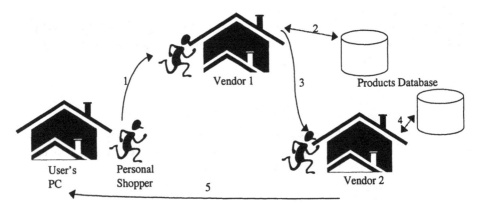

Fig. 7. Shopping scenario.

3-User installs the chosen component (Figure 8)

3.1 Notifying agent notifies users, places and device manager

3.2 Installer agent adds connection agent if needed   (e.g. Cellular phone
connection)

3.3 Component-Installer adds conversion agent if needed (e.g. text-voice, voice-
text,Ö.etc)

3.4 Installer updates device registry, adds device driver(s) through negotiations
with elements agents

3.5 Installer agent registers with the vendor

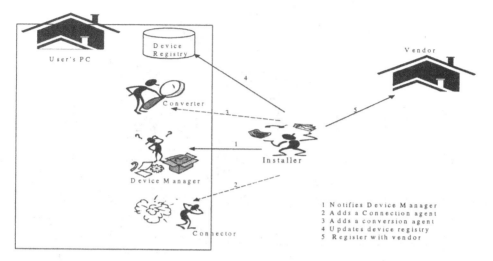

Fig. 8. Installing a new element (device).

5 First Generation Seamless Messaging Prototype in Agent Testbed

5.1 Seamless Messaging Application Requirements

Seamless Messaging must manage both asynchronous and isochronous (where immediate access to the user must be established) messaging, be easily customized by the end user, and be modular enough allowing features to be easily added or removed. For asynchronous messaging, a universal message box that can handle multi-modal messages is maintained. Thus, in a single unified view, the user can view a list of any incoming voice mail, email, fax or video messages. Voice mail and fax messages in the list show the number of the caller, their organisation (if caller id can pick it up) and an icon which can be selected to hear the voice mail or view the fax. Isochronous messaging requires immediate delivery of the message to the roaming user and the possibility of establishing a connection in the case of voice calls. Figure 9 provides a high level view and the flow of information between the seamless messaging agents.

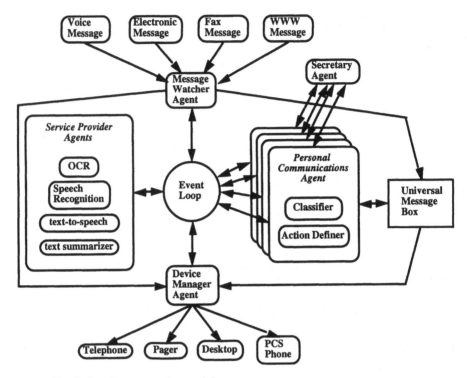

Fig. 9. Seamless messaging workflow.

Note that there are 2 paths for a message to take, either into the universal message box if it is asynchronous (connectionless) or directly connected through the Device Manager Agent to the output device for isochronous communications. The event loop is used to control the flow of communications between the agents.

The Seamless Messaging application is based on five sets of generic agent behaviours: the *Message Watcher*, the *Personal Communication Agent*, the *Secretary*, the *Service Provider* and the *Device Manager*. The *Message Watcher Agent* is designed as an event monitor that waits for incoming messages and formats these messages into a unified representation for the system to process. The *Personal Communication Agents* are designed to act based on a set of user specified constraints that the PCA interprets as rules at runtime. Note that the constraints are customizable at any point while the seamless messaging application is running. The user can even telephone the system and get into a dialogue with an automated speech interface to the PCA to modify their messaging requirements. The *Secretary* agent cooperates with the PCA to find the user to act on a message where the user has instructed the PCA action behaviour to ëcontact or find meí. The *Service Provider Agents* fulfil specialised behaviours in transforming messages from one form to another (for example, text-to-speech). Finally, the *Device Manager* Agent delivers the message by invoking its appropriate device driver behaviour and passing it the tailored information for delivery to the userís active device.

The *Event Loop* is used to illustrate that the agents communicate by placing inter-agent message blocks into a common area that other agents have access to. Message blocks contain an identification tag which triggers the appropriate recipient agent to take the message out of the common area and interpret it. The system is easily extendible by defining a new agent and the type of messages the agent should respond to. Although conceptually any agent can communicate with another on the *Event Loop*, in reality the workflow is constrained.

5.2   Personalizing the System

This particular system contains three personalisation functions: *Classifier and Action Definer behaviours* associated with the PCA, and a *Secretary* agent. The end user personalizes the system by specifying his/her desires to constrain the PCA and the Secretary agents.

5.2.1   PCA Classifier Behaviour

The *Classifier's* role is to look through an incoming message and classify the message into a set of classes defined by the user. The user can define any number and any type of classes for a message along with a set of rules that helps determine if the message fits the defined class type. These general structure of the rules is as follows:

```
DEFINE class_name AS
    ((attribute [operator] value)
    OR (attribute [operator] value).....
    AND ((attribute [operator] value)
    OR (attribute [operator] value)...)
    ...
```

In this example, the class_name will be associated with the messages that evaluate the expression after the keyword AS to TRUE. This expression is composed of a number of (attribute [operator] value) sets connected by Boolean operators. The currently available attributes are sender, recipients, number of recipients, date received, number of lines, subject, message body, and other defined classes. The operators available are =, !=,), >=, <, >. Obviously, not all are applicable to all fields, and the operator meaning can differ. For example, "number of recipients = 2" means an equality, whereas "body = 'this is an example'" means "body contains 'this is an example'".

5.2.2   PCA Action Definer Behaviour

The *Action Definer*'s role is to take a classified message and attach to that message an action to be performed on the message. Specification of the type of action to be performed to the message is similar to that done for the classifier agent, i.e. as a set of rules that map classes of messages to actions. A typical example of a rule has the following form:

IF class_name
        ACTION (action_1, action_2, ..., action_n).

Which specifies that if a message has been classified as class_name perform actions (action_1, action_2, ..., action_n).

Actions are divided into 2 types: direct and subjective actions. Direct actions are actions that can be resolved directly by the *Device Manager Agent* and require little or no interpretation. These actions can be serviced directly by the *Device Manager Agent*. Subjective actions require more interpretation and therefore are processed by the *Secretary Agent* before being serviced by the *Device Manager*. Examples of these two types of actions are the following, "phone_me '276- 2389'" is a typical direct action, while "contact_me" is a subjective action requiring further interpretation. These more abstract subjective actions are open to interpretation and are contextually interpreted based on the situated belief of the useri's availability.

5.2.3   Electronic Secretary Agent

The *Secretary* agent's role is to act upon any of the actions attached to a message. The *Secretary* must first decide the type of action the message has, is it a direct or a subjective action. If it is a direct action then it can be passed directly to the *Device Manager Agent* so it can process the message. If the message has the subjective action then the *Secretary* has to interpret the action. An example of this is the *contact_me* action.

The *contact_me* action requires the Secretary to use whatever resources are at its disposal to determine where the user may be. At the moment it uses a unique hierarchical scheduling structure to determine the availability of the user. The scheduling concept uses a three tier hierarchy representing three abstractions of time management. Commencing from a default description of a user's typical day (e.g. 9-5 Workday), to a more refined blocked-off scheduled events (e.g. 10-12 talk, 2-4 meeting), to finally a set of temporary changes which reflect the instant picture of where the user is located (e.g. 11-12:30 John's Office). The more detailed temporary changes take precedence over the user's scheduled events which in turn take precedence over the default day. Note that temporary events can be called and recorded by the PCA to dynamically alter the useri's availability.

5.3 Mediating the Devices: the Device Manager Agent

Devices are mediated using the *Device Manager* whose primary role is to determine the appropriate manner to deliver a message to a particular user device. This agent can use any of the available services from the *Service Provider Agents* to map a message to a particular format suitable for the device. Currently the system can call on two *Service Providers* for two types of mapping, a text- to-speech conversion (which is performed directly by the Computer Telephony server) and a text filter or summarizer trained on particular feature sets of a message. The later is used to interpret text messages and deliver them to a pager which is constrained by the number of text characters it can accept.

5.4 Implementation of Seamless Messaging

The system has been implemented using Lotus Notes as the underlying development environment [14;15]. The system, as currently implemented, can manage both voice and e- mail messages. Electronic mail messaging is a typical example of asynchronous communications and e-mails are therefore immediately stored into the Universal Message Box and processed. If a message is categorized to be immediately delivered to the end-user, the system takes the action to actively deliver the message through a pager, a telephone (using text to speech), fax, or by forwarding it to another e-mail address. Any message that is not categorized to be delivered to the user is kept in the Universal Message Box and can be retrieved, via the telephone or using an email client on a workstation. Voice calls are an example of isochronous communication and are not immediately forwarded to the Universal Message Box. Instead the incoming call is classified and delivered immediately to a device that can manage isochronous voice communications by connecting the calling party to the user in real-time. If this is not possible a voice message is taken and stored in the Universal Message Box. The voice message appears in the list of user messages and is manipulated in a manner similar to e-mail. The first generation has been actively in use as of 1996.

6 Future Work and Conclusions

Our future work will include the full implementation of the Network Management agents and the second generation seamless messaging prototype. The second prototype will introduce more complex behaviours to the PCA that would include reasoning based on more complex structures. This would allow the end user to state in simple terms more complex directives about managing their messages. We also recognise the need for a PCA firewall to protect it from being snooped on by other PCAs or by external Service Providers who may wish to target market messaging services based on user preferences. A first generation PCA launch tool has been partially implemented and is targeted for JAVA. We are examining the scalability issues of the applications and their related performance impact. The launch environment will be TINA-compliant since TINA seems to be one of the most promising candidates for future telecommunications architecture.

We believe Seamless Personal Information Networking presents a complex real-world problem to the agents community. We are developing a single unified testbed that incorporates diverse agent applications with shared behaviours and cooperative reasoning. Our emphasis is on the use of off the shelf computing and communications components in unison with the agent paradigm which provides an additional challenge to our work.

Acknowledgements

We would like to acknowledge the full SPIN team which includes Roger Impey and Ramiro Liscano who developed the first prototype in seamless messaging, Mansour Toloo and Larry Korba who are focussing on Network Management as well as Mark Vigder who is focussing on the Object-Oriented aspects of our agent work. We would also like to acknowledge our collaborators in the Mobile Agents Alliance at the University of Ottawa and Mitel Corporation. Finally, we would like to acknowledge Nortel Ltd. for their support with the wireless PCS 1900 BTS and their ATM Passport switch.

7 References

1. Abu-Hakima S., Liscano R., and Impey R.:Cooperative Agents that Adapt for Seamless Messaging in Heteregeneous Communication Networks. In IJCAI-96 Workshop on Intelligent Adaptive Agents, AAAI Press (August 1996) 94-103

2. Huhns, M.N. and Singh, M.P.:Cooperative Information Systems. In IJCAI-95 tutorial Notes, (August 1995).

3. Reinhardt , A.: The Network with Smarts. Byte, (October 1994) 51-64

4. Weihmayer, R. and Velthuijsen, H.: Application of Distributed AI and Cooperative Problem Solving to Telecommunications. In Proc. of the International Workshop on Distributed Artificial Intelligence, (July 1994)

5. Laufman, S.: The information marketplace: The challenge of information commerce, In Proc. of the Second International Conference on Cooperative Information Systems, (May, 1994) 147-157

6. Huhns , M.N., Jacobs, N., Ksiezyk T., Shen W., Singh M., and Cannata P.E.: Integrating Enterprise Information Models in Carnot. In Proceedings of the First International Conference on Intelligent and Cooperative Information Systems, (1993)

7. Rosenschein , J.S. and Zlotkin , G.: Designing Conventions for Automated Negotiation., AI Magazin, Vol 15, No. 3. (1994) 29-46

8. Gray , T., Peres , E., Pinard , D., Abu-Hakima , S., Diaz , A., and Ferguson , I.: A Multi-Agent Architecture for Enterprise Applications. In Working Notes AAAI-94 Workshop on Artificial Intelligence in Business Process Re- engineering, (July 1994) 65-72

9. Abu-Hakima, S., Ferguson , I., Stonelake , N., Bijman , E., and Deadman , R.: A Help Desk Application for Sharing Resources Across High Speed Networks Using a Multi-Agent architecture. In Proceedings of the workshop on AI in Distributed Information Networks, IJCAI-95, (August 1995) 1-9

10. Bradshaw, J.M., Dutfield, S., Benoit, P., and Woolley, J. D.: KAoS: Toward an Industrial-Strength Open Agent Architecture, Software Agents, Chapter 17. MIT Press (1997)

11. White , J.: Mobile Agents, Software Agents, Chapter 19. MIT Press (1997)

12. Park, A.: Multi-Agent Architecture Supporting Services Access, In Fifth International Workshop on Mobile Agents, (1997)

13. Zeid, Amir.: Towards Software Engineering for Mobile Agents, In Poster Session of the European Conference on Object-Oriented Programming, (1997)

14. Gordon, P.: Message Handling Application, NRC Work Term Report, (1996)

15. Yu Q.: Seamless Message Handling Application, NRC Work Term Report, (1996)

16. TINA-C Service Architecture, At www.tinac.com. (1997)

Agent Negotiation for Supporting Personal Mobility

A. Hooda[1], A. Karmouch[1], S. Abu-Hakima[2]

[1]University of Ottawa, Department of Electrical & Computer Engineering,161 Louis Pasteur
St. Ottawa, ON, Canada, K1N 6N5
amin@sol.genie.uottawa.ca, karmouch@elg.uottawa.ca
[2]National Research Council Canada, Institute for Information Technology, Seamless
Personal Information Networking Group, Ottawa, ON, Canada K1A 0R6
suhayya@ai.iit.nrc.ca

Abstract. Personal mobility – a fundamental characteristic of Nomadic
Computing - creates an environment in which a user roams without a wireless
laptop or a mobile phone and yet enjoys anytime, anywhere network
accessibility. This paper gives an overview of *Nomad's Personal Access System*
(*NPAS*), OSI application layer software that provides a personal mobility
environment within a virtual network. In addition, it focuses on the part of
NPAS, called *Site Profile Agent*. *Site Profile Agent* is an inter-site negotiating
agent that communicates with its peer to decide services for a nomad. This
work is motivated by the larger project of the *Mobile Agents Alliance*, a
collaboration that includes the National Research Council of Canada, the
University of Ottawa, and Mitel Corp.

1   Introduction

Nomadic Computing [2], a new mode in computing, spurred by the integration of the
data and telecommunication networks [3], known in North America as Personal
Communication Services (PCS) and in Europe as Universal Mobile
Telecommunication System (UMTS), offers an interesting application area for agent-
based design and programming. The nomadic environment, unlike its precursors, not
only provides efficient signal transmission to fixed and mobile devices, but also seeks
to provide anytime, anywhere, personalized information access to its end users
regardless of underlying network, device and location [2]. The instantaneous,
personalized information access requires software integration (also called middleware
in [2]) to provide a service-level interoperability among heterogeneous, self-contained
network infrastructures [2]. The ability of agents to communicate and co-operate
irrespective of the underlying programming language, transport protocol, etc. could be
exploited to achieve a high-level interoperability [1].   For example, agents
representing a nomadic user (user's preferences and needs), can negotiate with other
agents to dynamically configure the user's visited environment, an instance of which
is *Site Profile Agent*. It is an integral part of *Nomad's Personal Access System* (*NPAS*)
[5]. *Site Profile Agent* is an inter-site negotiating agent that communicates with its

peer, situated in a cooperative, virtual network of *NPAS*, to mutually decide computing and communications services for a nomadic user based on the user's personal profile. This paper presents an overview of *NPAS* and it particularly focuses on the design and implementation of *Site Profile Agent*.

NPAS, application layer software addresses *personal* mobility - a seminal characteristic of nomadic computing – which offers global computing and communications support to end users through wired or wireless devices available at any location. However, *NPAS* simulates a *personal* mobility environment within a private virtual network, spanning different organizational networks. It implements *dynamic mapping* between a user and the shared devices (or services) available at any location within the virtual network. The *dynamic mapping* of users to devices is attained using Internet's LDAP directory. It identifies the data management aspect of *NPAS*. For certain functions in data management, the agent concept is leveraged, which is the focus in this paper. Another, inherent aspect of *NPAS* is messaging that deals with the communication needs of mobile users. Messaging is largely addressed by the interaction of autonomous programs (agents) [6], representing users, services, and data resources. In the *NPAS* environment, a nomadic user is allowed to use shared device(s) in a new location, while away from his/her home or office desktop, fixed or mobile phone, etc. This capability allows personalization of network services in a visited location, which becomes a 'surrogate or virtual home' for a nomadic user.

1.1 Related Work

The existing wireless technology, being one of the enabling predecessors of PCS, provides a context to *dynamic mapping*. The basic requirement to find the location, be it of a device or a person, is an overlapping aspect in both *terminal* and *personal* mobility. Therefore, it is assumed that traditional *Location Management Techniques* (*LMTs*) will strongly influence the evolution of those for *personal* mobility [5].

Among other issues of *terminal* and *personal* mobility, the issue of *dynamic mapping* of users with changing devices has been highlighted in [2, 13]. A concrete solution to the problem of *dynamic mapping* is *NPAS*. Similar to *NPAS*, BERKOM II project [9] also deals with *personal mobility* and its ramifications (messaging services, which they categorize as service personalization and service interoperability). Their design approach draws on concepts from IN, UPT and mobile computing and on integration principles of TMN and their implementation is based on X.500/X.700, Electronic location Systems and TINA-C. However, they do not provide the inter-site negotiation for building service profile of a visiting user at a new location. *NPAS* is distinct because of its deliberations on the adaptability of traditional *LMTs* to manage *personal* mobility [5]. The strong recognition of the user, workspace and site-related data (defined later) as the basis for providing messaging services, also distinguishes *NPAS*.

The remainder of the paper is organized as follows: section 2 lists some assumptions underpinning the design of *NPAS*, section 3 gives an overview of *NPAS*, section 4 describes the design and implementation of *Site Profile Agent*, and section 5 presents future directions and conclusions.

2 Design and Messaging Assumptions

Today's telecommunications and data worlds (until the realization of PCS) are considered disconnected and self-contained. Therefore, the design approach for *NPAS* is incremental and does not assume replacement of or addition to the legacy user equipment. Thus, it does not assume that a nomadic user can be reached anywhere within its coverage by means of a personal number, which is the basis for UPT [3]. This means, for example, as usual a caller would dial a telephone number of the called party, and the call would be delivered the same way, as it does normally. However, an incoming call is intercepted at the home side of the called party, where the capability to intelligently forward the call to the called party's current location is provided. Intelligent forwarding is based on the users' preferences, messaging requirements, and visited site permissions, restrictions, etc. In case of an outbound call, initiated by a visiting user, device-based user identity authentication is required. *NPAS* assumes that if such authentication means are available, the visiting user's call (in a general sense user's actions) can be permitted or restrained based on the visited site's policies and the service profile of a visiting user.

At present, *NPAS* does not employ any automatic location tracking (e.g., active badges). Therefore, the onus is on a nomadic user to inform his/her location to network. In *NPAS*, a nomad subscribes at home with his/her unique id. In our case it's an e-mail address, which uniquely identifies a nomad at home and visited locations.

3 Design of Nomad's Personal Access System

Fig.1 below shows the architecture of *NPAS*. The key components are: (i) the *Data Repository (DR)*, which deals with the collection and maintenance of data for nomads and devices available in a workspace; (ii) the *Messaging Services (MS)*, uses *DR* information to render messaging services to nomads; (iii) the *Intermediate Objects (IOs)* is a middle-tier between the *Data Repository* and *Messaging Services*. In this work, it is strongly recognized that *Messaging Services* cannot be provided unless *DR* is in place. The three components are elaborated on below.

3.1 Data Repository

The *Data Repository* provides the functionality of the *dynamic mapping*. A nomadic worker while in transit is exposed to a spectrum of devices in the changing workspaces. To attain instantaneous access, independent of location and device, the network environment must adopt an efficient and scalable procedure to store and update the changing locations of nomads and the devices associated with those locations. As a first step, this requires liberating roaming users from fixed associations with devices by assigning each nomad a globally unique identifier. Subsequently, every nomad must either notify the network through an explicit logon or an implicit logon. An implicit logon is used in the sense that the nomad does not have to consciously inform the network about its whereabouts but the network through its sensors tracks the nomad's location [5].

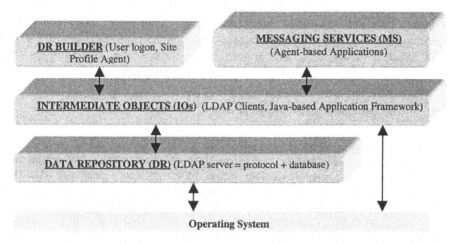

Fig. 1. Components of *NPAS*

After logon at a new location, the hosting network in view of the visitor's guest profile creates a dynamic association between a visiting user and shared devices available in that visited location. This is a passive association, which refers to a binding that is evaluated either at the reception of a message on the basis of the media and/or semantics of the message or at the origination of a call through device-based user authentication. This (non-time-critical) association also takes into account the integrity and security of the hosting environment, an important aspect of *NPAS* data management.

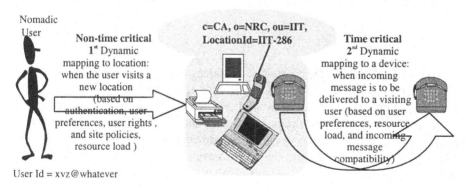

Fig. 2. A view of two-step dynamic mapping

At the time of an incoming call, the network does another dynamic mapping (which is time critical for synchronous communication), however, this time between a user and ultimate device, conforming to the media and/or of the message. This mapping can be termed as an active one, since it is used to 'deliver' the message, an essential goal of *NPAS* messaging. Fig.2 above shows an example of the workspace area: IIT-286 at NRC, Canada where a variety of devices is available for visiting users. The workspace area is statically populated in the *DR*, for more details see [5].

3.1.1 Implementation of Data Repository.

The *dynamic mapping* primarily requires a name binding and resolution service, e.g. a directory service. A directory service allows attribute-based search – that is both name and attribute can be used to lookup entries in the directory database. The *Data Repository* will be queried in situations where an attribute-based search is needed, for instance, when searching through a fixed-phone number to find a user's fax number. Since the Lightweight Directory Access Protocol (LDAP), the IETF directory standard [10] for a TCP/IP network, provides a directory service, which is distributed, scalable and highly protected, LDAP was selected to implement the distributed *Data Repository*. LDAP increasingly provides almost all ingredients of X.500 [11], though the distributed model is not fully-fledged as of LPAP v3 [15].

The types of the mobility-related data required to achieve the *dynamic mapping* are: (i) User-related data, which refers to the profile (the static data and learned data on user's location, preferences and messaging requirements) of the users at home and visited location; (ii) Workspace-related, which defines the types of devices that are available for visiting users in a visited workspace (location), and it also provides schema to store information on usage-load and health of the devices in that location area; and (iii) Site-related data, which contains information about the overall organization policies and the service contracts that are applicable to its known and new visitors. The detailed description of the user and workspace-related data is in [5].

3.2 Messaging Services

The *Messaging Services (MS)* constitutes messaging applications for (initially) voice, text, and (later) multimedia communications. Messaging applications are responsible for delivering incoming and outgoing messages by making appropriate interceptions. For instance: the acquisition of current location data on the nomad, conversion of media, and filtering a message based on the personal profile of a nomad. Some of these issues are also discussed in [6], however details are not provided. The capabilities required in such applications involve autonomous entities that can watch messages and trigger appropriate actions. Therefore, it is noted that the messaging applications of *MS* depend on the semantics of autonomous *agents*. The conception of network architecture for messaging applications, identification of concrete roles for agents to facilitate messaging and their strategic posting on the network distinguishes our work. Due to the limited space, these works cannot be presented here.

3.3   Intermediate Objects

The *IOs* is conceived from the implementation perspective. The first objective of *IOs* is to act as a middle-tier forming a traditional three-tiered architecture [14], widely used in web-based form processing. It gives a common channel for all *MS* agents/applications to access information collected in the *Data Repository (DR)*. The second objective of *IOs* is to provide an applications framework for building messaging applications.

3.4   NPAS's Location Management Technique

The *NPAS's LMT* is essentially based on the Home Location Register (HLR)/ Visitor Location Register (VLR) scheme that has been in use in IS-41and GSM. The objective of the *NPAS's LMT* is to find the current location of a roaming user and the devices available in that location. Each site maintains data on its *nmUser* (native mobile user*)*. The *nmUser* refers to a mobile user's profile stored at his/her home. The visiting mobile user submits his home e-mail address along with his/her identity authentication to home or visited site logon interface, as a first step. Based on this data, at most all enabled sites are searched. If the search is successful, then the negotiation is performed between the visited and home sites to determine the guest profile of the *vmUser*. This negotiation is performed by *Site Profile Agents* at home and visited sites. They dynamically assign fixed and mobile devices based on the several criteria, as described on below. As a final step, the negotiated guest profile is sent to the home site. If the result of the authentication search is negative, i.e., the visiting user does not exist in the distributed directory, it is assumed that the user is not a subscriber of *NPAS* and therefore logon is refused.

3.5   A 3-Site Virtual Network Scenario

This work is the part of the larger project undertaken by the *Mobile Agent Alliance (MAA)*, a collaboration that includes the National Research Council Canada, the Univ. of Ottawa, and Mitel Corp. The initial demonstration prototype of *NPAS* will be a virtual network of the three sites shown in fig.3 below. This testbed would allow the exploration of the practical issues of *personal* mobility. Fig.3 shows the three *Enabled Sites*. A site where *NPAS* is available is called here an *Enabled Site*. A set of *Enabled Site* forms an *Enabled Region*. A site where *NPAS* is not available is referred to as an *Un-Enabled Site*. Similarly, a set of *Un-Enabled Site* combine to form an *Un-Enabled Region*. Each *Enabled Site* maintains a Directory Information Base (DIB) for its organizational network. Formally, the directory's data model is based on a partial information strategy (nodes in the network contain database for the portion of the network) and each DIB is analogous to a "regional directory" of [12]. It should be noted in fig.3, that the three agents are posted to form a virtual network among the sites. These agents are representative of proxy [1] or interconnection agents that support the mobility of subscribers roaming through heterogeneous environments of these sites.

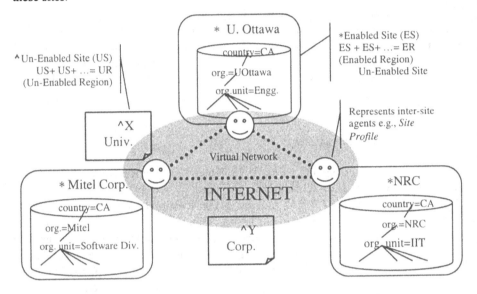

Fig. 3. A 3-Site Private Virtual Network on Internet

4 Site Profile Agent

Site Profile Agent is an inter-site negotiating agent that communicates with its peer to mutually decide computing and communications services for a roaming user. These agents are situated in a co-operative, virtual network. The agents negotiate solely one-on-one. The negotiation (i) respects the policies, security, and authorizations of the visited site; (ii) protects a visiting user's profile against snooping; and (iii) fulfills the requirements (communications and computing resources) of a visiting user for a virtual workspace at a visited site. *Site Profile Agent* only interacts on behalf of the user who is a subscriber of an *enabled site*. The sites involved in the user profile negotiation are assumed to have an a priori organization-level contract. This contract endorses the willingness of a site to be a 'surrogate home' for known visitors and explicitly specifies a set of pre-defined classes of services for those visitors. *Site Profile Agents'* negotiation only determines a resources (or services) profile for a visiting user. It has no influence in the establishment of any organization or visitor-level contract. The visitor-level contract identifies a visiting user who is not affiliated to an organization.

Site Profile Agent is not able to take input or feedback from the visiting mobile user. This enhances the security of the sites involved. To perform their roles autonomously, as representatives of a visiting user and a visited location, they depend solely on the user, location and site-related data stored in the corresponding *DRs*. If the negotiated profile seems unsatisfactory to the visiting mobile user or to his/her *User Agent* [6] (which knows about nomad's messaging criteria and performs the work of a personal secretary), it is assumed it would require human intervention. Such intervention comes from the user by either modifying their home *DR* or informing the visited site's *DR* administrator.

4.1 Inter-Agent Communication Using KQML

Analogous to objects in object-oriented model, agents' communication depends on the formalized message-based interfaces. However, these interfaces are application (or object) independent. Further, they are transparent to the underlying programming language, transport protocol, and operating system. They form a proposed a standard communication language for agent-based application and is known as KQML [1,8]. KQML messages are developed from the speech-act theory (a theoretical account of human communication). A KQML message contains speech-act (also called performative, with an understanding that it will result in action), which provides intention (attitude) of the message being sent. The pragmatic information (who is the sender and receiver, how to understand what is said, and how to identify a message received) and finally, an arbitrary content type, e.g., ASCII, and binary format, follow the speech act. The part of pragmatic information serves as a message-handling protocol.

Here, *Site Profile Agents* communicate using speech-acts and the message handling protocol of KQML. The scenario shown below in fig.5 uses three KQML performatives: ask-if, tell and untell. These performatives sufficiently support the discourse semantics required in the agents negotiation. For error and exception recovery, performatives in the intervention and mechanics category are used, however they are not illustrated here.

KQML does not commit to any interaction protocol: a sequence of message exchanged in any multi-message conversation. However, it does provide a set of primitive conversations, conversation policies, to validate the semantics of the performatives [8]. The conversation policies empower a KQML-speaking agent to select primitive conversations. The selection is based on the performative of the first incoming message. However, it does not give the ability to carry on a valid, arbitrary conversation. In applications, where an agent at the time of its creation is known to have a specific purpose, e.g., *Site Profile Agent's* conversation, it becomes redundant to put these conversation policies for semantic validation. Instead, where necessary, agent should be programmed in a way that if a need arises, it could select from the conversations provided to it [1].

4.2 Negotiation Between Site Profile Agents

Site Profile Agents' negotiation is defined off-line [1] as a pair of caller and callee conversations. The agent that initiates conversation is identified as the caller and the one that responds is ascribed as the callee. A conversation can be defined, using

Definite Clause Grammar [8], state transition diagram [1,7], etc. The conversations are represented here using finite state diagrams.

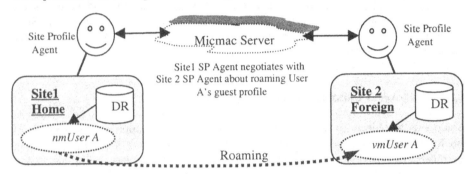

Fig. 4. Home and Foreign *Site Profile Agent* Negotiation

Fig.4 above, shows home-foreign *Site Profile Agent* communication through Mitel Corp.'s blackboard known as Micmac Server [16]. It is assumed the Site 1 and Site 2 are enabled locations and user 'A' is the *native mobile user* of Site 1(Site 1 is their home). The *NPAS* users logically have two options to logon. First, they can logon to Site 2 using the home site logon interface irrespective of whether they are physically at home or not. The second option, is to log on through the visited site logon interface irrespective of whether they are physically present at visited site or not. When, a user logs on to Site 2 through their home site logon interface, their home *Site Profile Agent*, as a caller, initiates negotiation addressing Site 2's *Site Profile Agent*. It is called here as a foreign *Site Profile Agent*, which becomes the callee in this negotiation. In the second case, a user logs on to Site 2 through the visited site logon interface. At this time, Site 2's *Site Profile Agent*, foreign *Site Profile Agent*, acts as a caller and initiates negotiation addressing their home *Site Profile Agent* at Site 1. Home *Site Profile Agent* at Site 1 now becomes the callee.

Consider an example that the user 'A' roams into Site 2, depicted as a dotted oval at Site 2 (fig.4) and assume that she logs on through the site logon interface associated to Site 2. Thus, foreign *Site Profile Agent* is invoked and it contacts her home *Site Profile Agent*. Finally, the output of this negotiation is either displayed at the time of logon or sent (by e-mail) for their information. Fig.5 below shows this conversation from the callee's perspective, which in this case is home *Site Profile Agent* at Site1.

Fig.5 is a simplified state diagram. It does not show details of the content sent through KQML messages. It also does not depict error or exception recovery rules arising from NACK or no response. Further, the language and ontology attributes of KQML messages are not shown.

In the first message of foreign *Site Profile Agent*, it informs home *Site Profile Agent* that they are going to start a conversation called 'Negotiate'. Within the same message, through the KQML performative 'ask-if', it requests agent authentication to home *Site Profile Agent*. Depending on the authentication id submitted it may produce ACK or NACK at home *Site Profile Agent*. If the outcome is ACK, it is sent to Site 2 using the KQML performative 'tell'. Then, the Site 2 sends the 'sign-up' data to the Site 1, which contains the logon information (e-mail id, user authentication, intended workspace area, etc.) of the user 'A'. This information could also produce ACK or NACK, subject to the correctness of the user identity data submitted during logon. When ACK is the outcome, then conversation goes into 'Profile Status', where it waits to receive one of the three results: 'NoProfile', 'Modify', or 'Oldfound', and the Services Option Tree, which is not shown in fig.5. 'NoProfile' indicates that Site 2 has no *vmUser* profile of user 'A' in its *DR*. 'Modify' indicates that Site 2 has found *vmUser* profile of user 'A' in its *DR*, however, is not able to activate the old profile. Perhaps due to the change in resource load, the site-related data of the Site 2, etc. Nevertheless, it is willing to negotiate. 'Oldfound' indicates that Site 2 has old *vmUser* profile of the user 'A' and is ready to activate it.

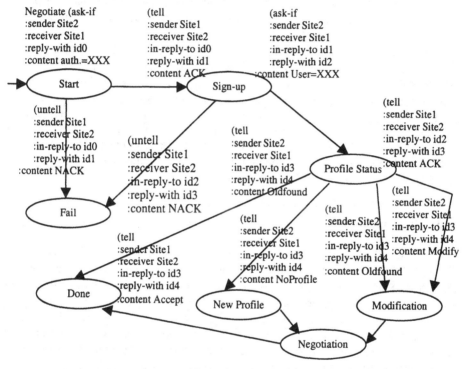

Fig. 5. (a) Callee Conversation (home *Site Profile Agent*-at Site1)

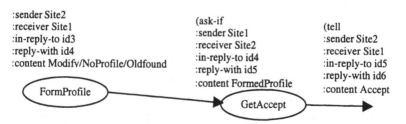

Fig. 5. (b) Sub-states of Negotiation State of fig.5 (a)

If 'NoProfile' is sent from Site 2, then Site 1 goes into 'New Profile' state and thereby it falls into the 'Negotiation' state. The 'Negotiation' state is decomposed in fig.5 (b) as two states: 'FormProfile' and 'GetAccept'. The 'FormProfile' state along with 'NoProfile' receives the Services Option Tree. The Services Option Tree is a tree-based data structure that stores any string data and their inter-dependence. The Services Option Tree contains information on what type of services are available at Site 2, what are their interdependencies, etc. The interdependence here refers to if one service is chosen from a branch then are there other services that can or cannot be chosen from the same level. Home *Site Profile Agent* traverses through the Services Option Tree and chooses the options desired in view of the user 'A' *nmUser* profile. Once the Services Option Tree is processed, it sends it back, as the 'FormedProfile', for the approval, and the Site 2 responds 'Accept', if satisfied. This brings the negotiation to the final state 'Done' in fig.5 (a). The Services Option Tree plays an important role in the agent negotiation, since it reduces the overhead of messages and thus greatly simplifies the conversation. However, this requires that foreign and home *Site Profile Agent* be complex enough to construct and process Services Option Tree.

In the case when 'Modify' is sent from foreign *Site Profile Agent*, the scenario remains the same as for 'NoProfile'. Finally, when 'Oldfound' is received from foreign *Site Profile Agent*, home *Site Profile Agent* has two options. It can choose to either send 'Accept' to end negotiation or go into 'Modification' state and follow through the same scenario as for 'NoProfile'.

4.3   Implementation

Fig.6 below illustrates components of *Site Profile Agent* and their functions. The *NPAS's LMT* is realized through the Manager Servlet, coded using the Java Servlet API from Sun Microsystems. The Java Servlet API is multithreaded; therefore, the conversations generated do not require asynchronous operation. When *Site Profile Agent* is acting as a caller, the Manager Servlet creates an instance of the caller conversation in a new thread. Also, there is server-like thread that runs continuously in the Communication Port to monitor a message posted in Micmac for a new Callee Conversation. In this case, it will pass the request to Manager Servlet to create an instance of a Callee Conversation. The Communication Port uses a public domain API provided by Mitel Corp.

5   Future Directions and Conclusion

Site Profile Agent negotiation is modeled as a single, large conversation, which could have been avoided by decomposing it into sub-conversations. This approach is more effective as it allows reuse of primitive conversations. For example, *Site Profile Agent's* meta-conversation could plug-in agent's mutual authentication conversation into the conversation 'Negotiate'. At present, a simple list is being used in place of the

Service Option Tree, but work is in progress to develop the representation and processing of the Service Option Tree. Such a tree could include prices for the services to define the billing, as well. The Negotiation protocol is also being reviewed to address error recovery.

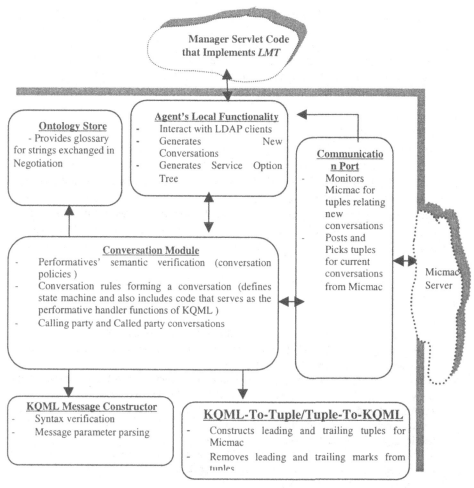

Fig. 6. Components of *Site Profile Agent*

This work explores the application of agents in *personal* mobility data management. Particularly, it discusses *Site Profile Agent* that achieves dynamic creation of visiting user profile in view of the visited site policies and native mobile user profile. For achieving this task, it employs inter-site agent negotiation using KQML-based conversations. The significant part of this negotiation is made possible due to the LDAP-based *Data Repository (DR)* and the conceptualization of a data structure, called the Service Option Tree. The Service Option Tree stores the services offered at visited site and their interdependencies. The Service Option Tree provides an efficient tool to reduce the time and bandwidth-consuming message passing, and simplifies the complexity of the negotiation.

The design and implementation of the messaging applications is underway. The *Data Repository* and *Site Profile Agent* is under review for the enhancement of mobility-related data (including site policies) and plug-in conversations, respectively. The site logon user interface is developed as a Java Applet. The platform used for software development is Windows NT4.0. So far, the coding is done using JDK1.1.3-5, Java Servlet API, Java LDAP API 1.0 beta, and Java Micmac API. Netscape's Directory Server 3.0 b2 is being used as a stand-alone LDAP server, where the user and workspace-related X.500 object classes are stored and the sample data is stored. Further, Java Web Server 1.0.3 is used for Servlet execution environment

Acknowledgement

The *Telecommunication Research Institute of Ontario* and *Mobile Agents Alliance* have supported this work. Dr. Mark Vigder of the Software Engineering group at NRC was especially helpful in critiquing the implementation of this work. The state diagrams have been much improved with his help. Finally, we would like to thank to all the researchers in the SPIN team for their support and many thoughtful, informal discussions.

6 References

1. J. Bradsahaw (ed.), Software Agents, AAAIPress/The MIT Press, 1997
2. R. Bagrodia, *et al.*, "Vision, Issues and Architecture for Nomadic Computing", IEEE Personal Communications, 2(6), pp.14-27, Dec. 1995
3. O. Spaniol, *et al.*, "Impacts of Mobility on Telecommunication and Data Communication Networks", IEEE Personal Communications, 2(5), pp. 20-33, Oct. 1995
4. Thomas Magedanz, "On the Impacts of Intelligent Agent Concepts on Future Telecommunication Environments", Proc. of 3rd Int. Conf. On Intelligence in Broadband Services and Networks, Germany, 1995, 396-414
5. A. Hooda, A. Karmouch, S. Abu-Hakima, "Personal Mobility Management Using LDAP Distributed Directory and Static Agents", to appear in 5th Int. Conf. on Intelligence in Networks, May 10-15, 1998, France
6. S. Abu-Hakima, *et al.*, "A Multi-Agent System for Seamless Messaging by Email, Fax or Voice Mail", Proc. of Int. Joint Conf. on AI, IJCAI'97 Workshop on AI in Distributed Information Networking, Negoya, Japan, Aug. 24, 1997, reference w4 to IJCAI, pp.9-16. http://www.nrc.ca/iit/SPIN_public, 1997
7. B. Falchuk, A. Karmouch, "The Mobile Agent Paradigm Meets Digital Document Technology - Designing For Autonomous Media Collection", to appear in, Multimedia Tools and Applications, Kluwer Academic Publishers, 1998
8. Y. Labrou, "Semantics for an Agent Communication Language", Ph.D Thesis, Computer Science Department, University of Maryland, Baltimore County, 1996
9. T. Eckardt, *et al.*, "A Personal Communication Support System based on X.500 and X.700 standards", Computer Communications, 20(3), pp. 145-156, May 1997

10. W. Yeong, T. Howes, S. Kille, "Lightweight Directory Access Protocol," RFC 1777, March 1995. W. Yeong, T. Howes, and S. Kille, "Lightweight Directory Access Protocol", RFC 1777, ftp://ds.internic.net/rfc/rfc1777.txt , Mar. 1995

11. ITU-T Recommendation X.500 (1993) | ISO/IEC 9594-1 :1993, Information technology – Open Systems Interconnection, "The Directory: Overview of Concepts, Models, Services", http://www.dante.net/np/ds/osi.html, 1993– Open Systems Interconnection – Web Site - http://www.dante.net/np/ds/osi.html

12. B. Awerbuch and D. Peleg, "Online Tracking of Mobile Host", JACM, 42(5), pp.1021-1058, Sep. 1995.

13. C.G. Harrison, "Smart Networks and Intelligent Agents", Proc. of Mediacom'95, Southampton, UK, Apr. 11, 1995

14. R.Orfali, *et al.*, "Client/Server Programming with Java and CORBA", John Wiley & Sons, Inc., 1997

15. M. Wahl, T. Howes, S. Kille, "Lightweight Directory Access Protocol (v3)," RFC 2251 December 1997 M. Wahl, T. Howes, and S. Kille, "Lightweight Directory Access Protocol (v3)", RFC 2251, ftp://ds.internic.net/rfc/rfc2251.txt, Dec. 1997

16. Micmac WebSite, http://micmac.mitel.com.

Development of a Multi-agent System for Cooperative Work with Network Negotiation Capabilities

Francisco Garijo[1], Juan Tous[1], José M. Matias[1] Stephen Corley[2], Marius Tesselaar[3]

[1] Telefónica I+D, Emilio Vargas, 6, E-28043 Madrid, Spain.
e-mail: fgarijo@tid.es

[2] BT Laboratories, Martlesham Heath, Ipswich, UK. IP5 3RE.
scorley@everest.srd.bt.co.uk

[3] KPN Research, P. O. Box 15000 ,9700 CD Groningen, The Netherlands.
M.J.A.Tesselaar@research.kpn.comv

Abstract. This paper describes the architecture and operation of a Multi Agent system for providing end users with an intelligent interface for video conference and cooperative work services. The system consists of negotiating Agents: the Personal Communication Agents offers the user an intelligent interface to the service, it negotiates the best conditions in terms of quality of service and costs, the Service Provider Agent (SPA) supports the provisioning of telecommunication services to customers; the Network Provisioning Agent (NPA) provides network connectivity upon requests from the SPA. The physical setting is made up network nodes interconnected through two Public Networks; the Internet and ISDN. Network nodes are based on PCs running Windows 95 /windows NT, and Unix work stations. The system is being developed as part of the EURESCOM P712 project. The objective of the project is to assess and make recommendations on the applicability of intelligent and mobile agent technology to telecommunications service and network management. The evaluation criteria and the current status of the system are presented in the paper.

1   Introduction

Recent progress in   network infrastructure and distributed processing has made possible the marketing of new telematic services incorporating "intelligent features" according to user demands. Agent oriented technology might be an important source of technical solutions for modeling and implementing   those services. Existing agent based services on the Web are useful examples showing the applicability of agent technology for the design and implementation of commercial services. Nevertheless, combining agent technology - concepts, techniques, methods and tools- with existing service and software technology gives rise to a number of open issues that need careful evaluation. The meaning of terms like 'agent', 'intelligent agent', 'agent based architectures', 'agent development', etc., have been widely discussed [1] [5] [13][21]. Other engineering issues such as, agent specification, agent communication, agent engineering, agent development tools, etc., have also been presented [10][13][20]. Unfortunately the advantages of agent based proposals, over existing protocol specification and implementation, and other distributed processing solutions are not clear. From an industrial point of view the potential, of agent oriented technology needs to be demonstrated and evaluated in practice with prototype applications, in order to asses its maturity and the possible risks. This is the main objective of the EURESCOM (European Institute for Research and Strategic Studies in Telecommunications) Project P712 "Intelligent and mobile Agents and their applicability to service and network management". The evaluation approach taken by the project consists on building prototype systems using agent based solutions. Evaluation data will be gathered by carrying out a number of experiments directly on the prototypes. In addition, equally valuable data will be provided by the experience of building the prototypes. Two cases study have been identified in the areas of maintenance and dynamic connection management. Each case study defines the functionality to be implemented, the scenarios to be demonstrated and the experiments to be carried out. This paper describes the configuration case study covering dynamic connection management. The Multi Agent System provides multimedia meeting services to mobile users traveling around the world. The connections needed for service provision are negotiated and allocated dynamically according to users needs. The experimental Multi Agent setting, and its functionality is described in section 2. including agent interaction and the agent environment. Agent's design is presented in section 3. It is based on distributed object oriented principles, incorporating the session concept for peer to peer agent communication. The evaluation parameters are described in section 4, and section 5 draws the conclusions and the current status of the project.

2 Configuration Case Study

The experimental Multi Agent setting is based upon the scenario described in FIPA Part 7 [9] which focuses on the agents negotiation capabilities for dynamic connection management. The Multi Agents' System provides multimedia meeting services to end customers traveling to different locations and wanting to contact other colleagues in different cities around the world. The user connects his/her portable PC to the hotel's local telecommunication resources, and activates the Personal Communications Agent (PCA). When the PCA is activated, it first registers its location to the Local Registration Authority Agent (LRA). The user could then ask the PCA to establish the multimedia meeting within 5 minutes, for the duration of approximately 2 hours, and within a budget of $500."

The agent asks the LRA for the addresses of Service Provider Agents (SPAs), and then starts to contact them to see what possibilities are available. The SPAs will then contact different Network Provider Agents (NPAs) to see what offers are available to set-up network connections. Once the SPAs have found suitable NPAs that can provide the service, they will make provisional bookings and report back to the PCA. The PCA will then select the most suitable SPA to provide the service. The contract is now 'signed' between PCA and SPA, and the network connection is activated by the SPA. The SPA will then convert the provisional booking into 'contracts' with the different NPAs.

During the lifetime of the service, the PCA will be actively monitoring the fulfilment of the contract and log any deviations.

The agents involved in the service and the network setting are represented in figure 1

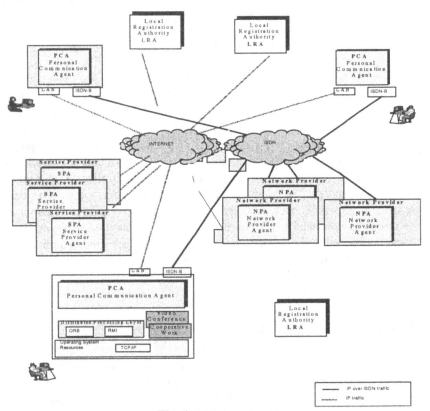

Fig. 1. Multiagent Setting

2.1 Agent Environment

The physical setting is made up network nodes interconnected through two Public Networks; the Internet and ISDN. Network nodes are based on PCs running Windows 95 /windows NT, and Unix work stations. The PCA communicates with other agents using the facilities of the Distributed Processing Environment (DPE) (ORB, RMI, etc.). The DPE protocols are based on TCP/IP. TCP/IP over Internet connections will be used for the agent' dialogues, while ISDN will be used to set up the connections needed to support video conference and cooperative work.

Collaboration with EURESCOM P715 has been established to enable the use of the ISDN European Services Platform (ESP) that is available between the project partners. The ESP is used to demonstrate provisioning of ISDN services using CORBA.

The software agent platform is based on Java and the Voyager platform version 2.0. [21]. PCA Inference capabilities are based on Ilog Rules Java [12]. The agents will migrate between the different partners over the Internet using the Voyager mobility services. This is particularly useful for the SPA that will travel between different locations to ease the negotiation process.

2.2 Agent Functionality

2.2.1 Personal Communications Agent (PCA)

The PCA represents the customer in it's dealings with Service Providers. The key functions performed by the PCA during service provisioning are as follows :

- Elicit and validate local telecommunication service resources such as- connection points, terminals, etc. -.
- Elicit and validate service participants' destination and service constraints for example, - starting time, duration, cost, etc.-
- Negotiate with Service Providers Agents (SPAs) the best cost-benefit balance for service provisioning in terms of the constraints and preferences defined by the user.
- Provide the user with monitoring capabilities to supervise the negotiation process, or just wait for a message to initiate the service.
- Enhance user profile through dialogue input. The PCA should use the user profile to minimize the dialogue with the user, asking questions like "would you contact the same people as you did the last time you where here?".
- Explanation capability. Inform the user why the service cannot be established. If some of the partners cannot be reached, the PCA should also inform the user , asking whether the service should continue or stop.
- Exhibit learning capability based on the classification of previous cases. Classification criteria are based on users preferences, list of partners, user locations, etc.

The PCA works autonomously on behalf of the user for service negotiation and service provisioning. User oriented functionality and negotiation facilities involves collaborative dialogues; PCA to User, and PCA to SPA.

2.2.2 PCA Interactions

User-PCA interaction. The dialogue between the user and the PCA is performed through the computer screen using visualization windows based on text, menus icons, active buttons, etc. The PCA should provide a robust user-friendly interface facilitating:
- Agent utilization: Configuration, activation, end, start, stop, monitoring.
- Acquisition and validation of service data: Service participants, service scheduling service cost, etc.
- Service control: Service activation, service interruption, cancel, re-start, etc.
- Service monitoring.

PCA-SPA Interaction. The PCA aims to negotiate with SPAs the best price/quality ratio for the service required by the user. PCA-SPA negotiation conversations are made up of messages sequences to:
- Communicate service requests.
- Get service offers.
- Communicate the selected SPA.
- Accept a service contract.
- Accept counter-offers.
- Execute contract.
- Resume contract.

PCA-LRA Interaction. The LRA manages Agent localization data. The PCA interacts with the LRA to:
- Register when it changes its location.
- Get the list of service providers.

2.2.3   Service Provider Agent (SPA).

Each Service Provider Agent represents the interests of a telecommunications service provider and supports the provisioning of telecommunication services to customers. It adopts two distinct roles :
-Client of network services offered by NPA
-Provider of a variety of telecommunication services to end customers.
The key functions performed by the SPA during service provisioning are:
- Authenticate the user.
- Determine component software/network service requirements.
- Negotiate with the PCA, terms and conditions of the delivery of the service.
- Identify secure Network Provider Agents (NPAs) for component services.
- Negotiate with NPAs for component network services specifying quality of service, bandwidth, source, sink(s), etc. The SPA tries to find the optimal solution in terms of quality of service and cost for providing the service to the end-user (through the PCA).

The SPA communicates with the PCA and with the NPA.

2.2.4   Service Provider Agent Interactions
SPA-PCA Interaction. The SPA communicates with PCAs to:
- Negotiate service contracts.
- Start contracted services.
- Invite parties involved in contracted services to participate.
- Provides the configuration services upon requests from PCAs.

SPA-NPA Interaction. The SPA uses the NPA to provide the PCA with the network connections needed for the requested service. The interaction consists of:
- Receiving requests from a PCA
- Translating and relaying the requests into call-for-proposals to one or more NPAs
- Collecting the responses from NPAs
- Negotiating contracts with NPAs
- Requesting the network connectivity
- Activating services
- Terminating the network connectivity after the termination of the configuration service.

2.2.5   Network Provider Agent (NPA).
The Network Provisioning Agent (NPA) represents a network domain. Its major responsibility is the provisioning of network connectivity upon requests from the SPA.. For this purpose, the NPA has to interact with the SPA representing the customer, the network management system representing the local network domain and with other NPAs representing other network domains in the global environment.

The key functions performed by the SPA during service provisioning are as follows:
- Negotiate with the PCA, terms and conditions of the delivery of the service.
- Negotiate with other NPAs, terms and conditions of external connection segments.

To provide the requested connection, the NPA will have to first break down the task into local and external connection segments, based on some service strategy and knowledge about the global network environment. Then it will try to reserve connection segments in its local domain and request other NPAs to reserve segments in their domains in order to connect the sources and destinations

2.2.6   Network Provider Agent Interactions
NPA-SPA Interaction
The NPA offers the network connection service to the SPAs via the following capabilities:
Processing the call-for-proposals from the SPAs.
Generating bids to the SPAs.
Providing network connectivity upon setup request from SPA.

Updating the network connectivity following the requests from SPA.

Releasing the network connectivity after the termination of the configuration service.

NPA-NPA Interaction

Other NPAs offer network connection service to the requesting NPA.

At this moment, we propose to adopt the same interaction protocol as NPA-SPA and SPA-NPA for NPA to NPA' interaction. This will be refined in the future after identifying the specific interaction.

2.2.7   Local Registration Authority (LRA)

The LRA plays the role of service broker. It provides service providers with information and users' location data. Interactions between agents and LRA follow the client-server model. The key functions performed by the LRA during service provisioning are :

– SPA registration.
– PCA registration when it changes its location.
– Provision of the list of service providers.
– Provision of location information of service participants' PCAs.

3   Prototype Design

The configuration prototype design is defined using an OO approach see figure 2-.

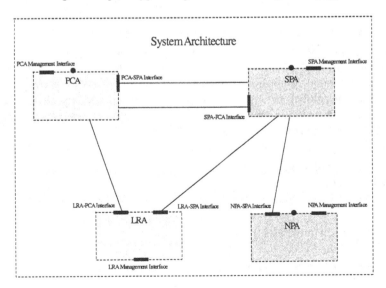

Fig. 2. System Design architecture

At the highest level the prototype is defined as a system formed by several subsystems: PCA, SPA, NPA, and LRA. There will be more than one instance of PCA, SPA and NPA. Each agent provides two interface types. The management interface contains the opertations needed for agent management such as agent monitoring, agent configuration and agent control. Agent to agent interfaces contain the operations needed to perform peer to peer communication.

Table 1 provides a summary of the operations included in each agent-agent interface.

Interface Operation	Client	Description
Interface PCA		
AceptOffer(bid_ref#,offer)	SPA	SPA send offers for a bid proposal
AceptContract(offer_ref#)	SPA	SPA send contract for a service
ServiceInvitationRequest(ref#, service type, quality parameters, calling party)	SPA	SPA invites a PCA to join in a service requested by another PCA
Begin_Service(contract#, participants' IP number list, service parameters)	SPA	activates the selected service contract and configures user's hardware
Interface SPA		
Bid_Request(ref#, service type, quality parameters, offered price, bid validity time, user list)	PCA	Ask a SPA for an offer for a specified service
Contract_Request(offer_ref #)	PCA	Ask a SPA for a contract for a specified service
Activate_Service_Request (contract#)	PCA	Requests the SPA to start a service
ConfirmService(contract#)	PCA	Ask a SPA for a service confirmation
CancelService	PCA	Ask a SPA to cancel a service
End_service(contract #)	PCA	Ask a SPA to end a service
Interface NPA		
ConfigureService(CID, QoS)	PCA	ask NPA if a connection necessary to deliver a service to PCA can be fulfilled
TearDownService(CID, QoS)	PCA	Teardown the connection
ConfigureConnection(QoS, CID)	PCA	Activate connection with QoS
ConfirmConnection(CID)	PCA	NPA can provide desired connection.
Interface LRA		
Register(agent_id)	PCA	PCA registers itself with a LRA
Deregister(agent_id)	PCA	PCA deregisters itself with the LRA
GetSPAList()	PCA	Get a list of known or available SPAs
GetUserLocation(user)	*SPA,PCA*	Get the location of the user

3.1   PCA Architecture and components

The PCA architecture -figure 3- uses the TINA session concept [20] to achieve parallel interactions among different subsystems. It consists of the following subsystems.

Agent Management. Its goal is to achieve the functionality associated with the management interface. It will not be implemented in the case study

Task Manager. The Task Manager controls the PCA behaviour. It processes the inputs received through the Management interface and the PCA-SPA interface, tracking the global agent state, and monitoring the behaviour of the PCA. It also creates, monitors, and co-ordinates the sessions needed for

- User-PCA dialogue (User Session Manager).
- PCA-SPA dialogue (Service Session).
- Enhancing the user profile according to the service user (User Profile Manager).

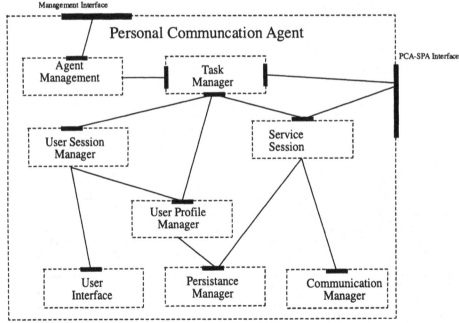

Fig. 3. PCA Architecture

User Session Manager. The User Session Manager processes the user inputs through the user interface. It manages and controls the dialogue with the user.

Service Session. The Service Session holds the computational resources needed to obtain the service functionality required by the user. The Service Session, controls the negotiation process with the SPAs, and set up   the communication resources needed for the provision of the service. The Service Session uses the Persistence Manager to stored its internal state.

User Profile Manager. The User Profile Manager (UPM) receives inputs from the Service Session to obtain user profiles. When the user sessions terminate, the UPM updates the user profile according to the service results.

User Interface. Represents the computational and graphical resources needed for user interaction

Persistence Manager. Provides a persistence services to the PCA

Communication Manager. Provides an interface to use and control the communication resources needed for service provision.

The **Task Manager** internal structure is in figure.4. The **Task Control Manager** (TCM), is represented by a symbolic processing module made up of Task Control Knowledge defined by means of plan libraries and an inference engine.

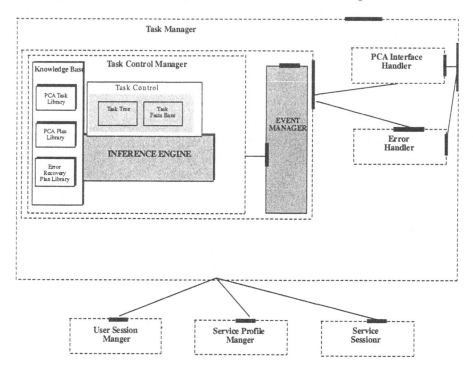

Fig. 4. Task Manager Architecture

The TCM receives events through the Event Manager. The reasoning engine uses the events to select plans from the plan library, creating a goal tree, and selecting an appropriate task to solve the pending goals. Task execution allows the control of the external resources asynchronously.

214

The User Session Manager (USM) creates and controls user sessions via the User Interface -Figure 5- The symbolic processing module contains the knowledge needed to guide and control the user-PCA interaction. USM KB is made up of service objects, information acquisition plans, input validation plans, and task resolution methods. The working memory is structured into two goal spaces: Service Monitoring and Control workspace keep track of the user input related to service monitoring , and Service definition control deals with Service definition and validation.. The USM uses the User Profile Manager to get the user profile when new service sessions start, and storing enhanced version of the user profile when service sessions terminate.

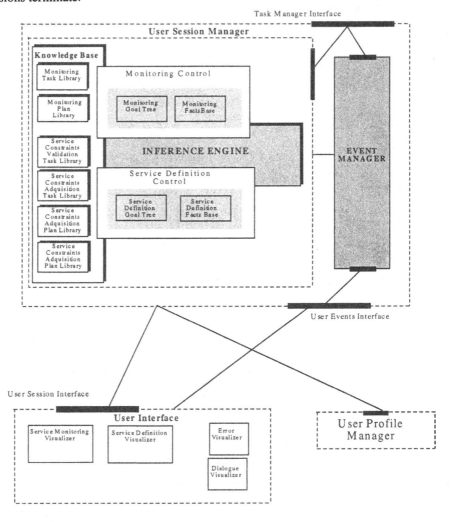

Fig. 5. User Session Manager Architecture

The structure of the Service Session is in shown in figure 6. The Service Session Control module creates, activates and monitors the Negotiation Session (NS), the User Service Participation Session (USPS) and the Service Control Session Resource (SCSR). The NS performs the negotiation process with SPAs. The USPS achieves service control, and the SRCS provides the interfaces needed for resource management

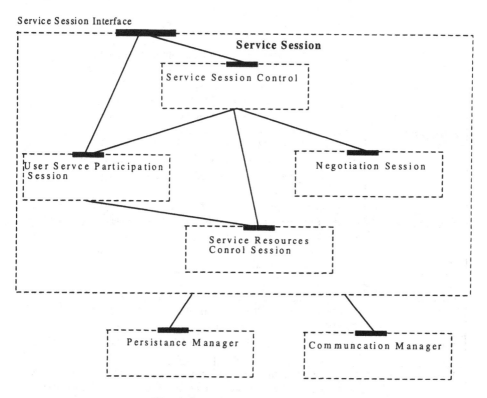

Fig. 6. Service Session Architecture

User Profile Manager. Its internal structure is based on a symbolic processor with specific knowledge to classify the data obtained from the SS. This new information is included in a new enhanced version of the user profile. The SS will use the enhanced version to improve the dialogue with the user in future service demand to the PCA.

4 Experimentation and Evaluation

The metrics and the experiments to be carried out with the prototype are described in the following table. The experiments describe those aspects of agent operation that are to be evaluated and the metrics that need to be acquired in order to understand the use of agent technology.

The information obtained will be used to assess the applicability of agent' technology and to draw recommendations for its utilization in network and service management.

Metric	Experiments
Mobility aspects	Mobility aspects will be measured for the Personal Communications Agent and the Service Provider Agents. Agent platforms will have a time-stamping mechanism in the AMS that will leave time stamps when registering, deregistering etc. The information is sent in messages to the home agent platform where a test agent will further process the timing information.
Communication capabilities	Communication capabilities will be measured / characterized as applicable between the following agents: PCA – SPA and SPA - NPA Qualitative metrics are obtained by observation of the agent. Quantitative metrics are obtained using a measuring and counting mechanism that will: • Measure communication time in seconds as the end-to-end time for each message to be transmitted and processed by agents. • Number of messages per communication phase. • Complexity of problem as the number of steps required to solve the communication problem.
Life cycle rela capabilities	Creation time, will be measured for: The PCA that travels to agent platforms of service provider agents. Destruction time, will be measured for: The PCA after the connection has been terminated.
Security	Security issues will be identified for all agents which need to register with a specific agent platform.
FIPA conformanc	The overall prototype will be assessed according to the FIPA conformance checklist (assuming one exists by the time we come to complete the experiments).
Business benefit	Business benefit comparisons will be made between our agent solutions and conventional solutions for the following activities like: • manual setup of connections between parties • manual maintenance of the connections • manual tear down of the connections

5 Current Status and Further Work

The project started in March 1997 and will end in March 1999. After an initial domain analysis activity the case studies were agreed and defined. Prototyping started in late 1997 and will continue until end July 1998. Two engineering approaches have been addressed for prototype implementation. One approach follows FIPA recommendations for agent specification and implementation. The second one, representing Telefonica's view in the project, is based on Open Distributed Processing Recommendation and Object Oriented Modeling. The Multiagent Architecture is modeled as a collection of Distributed Systems offering several interface types.

The justification of this approach follows from several considerations. Building commercial services will require agent technology to be integrated with conventional software technology, data base technology, and distributed processing. As existing methodologies, computational environments and development tools, allow this integration, it is important to carry out experiments to assess the development of systems satisfying the requirements and the functionality required for Multi Agent Systems. Most of the new engineering concepts defined by OMG and TINA seems applicable to agent based systems. Unfortunately people developing conventional telecommunication services are not familiar with symbolic techniques. A pragmatic approach to introduce these techniques is to design the symbolic component as a conventional component, hiding the symbolic implementation, and avoiding side effects with the rest of the subsystems. This is a way to demonstrate in practice the need of cooperation between symbolic processing components, and imperative components, to provide enhanced functionality to end customers. On the other hand, new ODP concepts like the TINA session concept, are better suited to the specification and implementation of peer to peer conversations among computational entities, rather than conventional message based approaches. The TINA session facilitates dialogue specification, avoiding low level message handling and correlation during the implementation.

The evaluation experiments will provide the data needed to analyze the advantages and disadvantages of FIPA based approaches as well as ODP approaches. As prototype implementation is planned to be finish by the end of July, initial results about implementation issues and system operation might be reported and discussed in the workshop.

Acknowledgements

The authors would like to thank the EURESCOM organisation, particularly Juan Siles, and other members of the Project team for their support and help in producing this paper.

Disclaimer

This document is based on results achieved in a EURESCOM Project. It is not a document approved by EURESCOM, and may not reflect the technical position of all the EURESCOM Shareholders. The contents and the specifications given in this document may be subject to further changes without prior notification. Neither the Project participants nor EURESCOM warrant that the information contained in the report is capable of use, or that use of the information is free from risk, and accept neither liability for loss or damage suffered by any person using this information nor for any damage which may be caused by the modification of a specification. This document contains material which is the copyright of some EURESCOM Project Participants and may not be reproduced or copied without permission. The commercial use of any information contained in this document may require a license from the proprietor of that information.

6   References

[1] J. Bradsahaw, "Software Agents", : AAAIPress/The MIT Press, 1997

[2] Davis and Smith, R.G (1983). Negotiation as a Metaphor for Distributed Problem Solving. Artificial Intelligence 20(1): 60-109

[3] B.Chaib-draa. Industrial Applications of Distributed A.I.. Communications of the ACM, Vol 38, N.11, pp:49-53, Nov. 1995

[4] Y. Demanzeau. From Interactions to Collective Behavior in Agent-Based Systems. 1st European Conference on Cognitive Science, Saint-Malo, France, avril 1995

[5] M.R.Genesereth, S.P.Ketchpel. Software Agents. Communications of the ACM. Vol 37, 7. pp:49-53 July 94

[6]EURESCOM P712 web page http://www.eurescom.de/public/projects/P700-series/P712.HTM

[7] S. Hedberg Agents for sale: firs wave of intelligent agents go commercial. IEEE Expert Dec 1996 pp 16-23

[8] FIPA Web Page, http://drogo.cselt.it/fipa

[9] FIPA. Application Design Test: Personal Assistant. FIPA´97 Draft Specification: Part 5. Revision 2.0.P. Kearney, D. Steiner, editors.1997

[10]Garijo F.J., Hoffman D. A Multi-Agent Blackboard Based Architecture for Operation and Maintennance of Telecommunication Networks. Proc of Avignon 1992 pp 427-438. Jun 1992

[11] Hylton, Jeremy, Ken Manheimer, Fred Drake, Roger Masse, Guido van Rossum, and Barry Warsaw "Knowbot Programming - System Support for Mobile Agents", Sept, 1996

[12] Ilog Rules http:// WWW.Ilog.com

[13] Maes, P., "Agents that reduce work and information overload", Communications of the ACM, 37(7): 31-41, July 1994

[14] Object Management Group, "http://www.omg.org/".

[15] C. J. Petrie Agent Based Engineering, the Web, and Intelligence. IEEE Expert Dec 1996 pp 24-29

[16]Pruitt, Dean G (1981). Negotiation Behaviour. Academic Press.

[17] K. Sycara, A. Pannu, M. Williamson, D.Zeng, K Decker. Distributed Intelligent Agents. IEEE Expert Dec. 1996pp36-45

[18]Sycara, K.P (1989a). Argumentation: Planning Other Agents' Plans. Proc. of IJCAI, Detroit, pp 517 - 523.

[19] Velthuijsen, H. and Griffeth, N., "Negotiation in Telecommunications Systems", In Notes AAAI Workshop on Cooperation among Heterogeneous Intelligent Systems, pages 138-147, San Jose, CA, 1992

[20]Telecommunications Information Network Architecture, "http://www.tinac.com".

[21] Tina7 C.Abarca, P.Farley, J.Forslow,J.Garcia. T. Hamada, P.F.Hansen, S.Hogg, H.Kamata, L.Kristiansen, C.A. Licciardi, H.Mulder, E.Utsunomiya, M.Yates. Service Architecture V 0.4 April 1997 TINA-C Deliverable

[22] Voyager Platform http:// WWW.objectspace.com/voyager

[23] K.Urzelai, F.J.Garijo. MAKILA: A Tool for the Development of Cooperative Societies. Lecture Notes in Artificial Intelligence 830 (Proc. of MAAMAW'92). pp: 311-323,1994.

[24] White, J. Mobile Agents White Paper. General Magic Inc. http://www.genmagic.com/agents/Whitepaper/whitepaper.html

[24] M. Wooldridge, N.R. Jennings. Intelligent Agents: Theory and Practice. Knowledge Engineering Review, 10(2), 115-152, 1995

From Interoperability to Cooperation: Building Intelligent Agents on Middleware

Bruno Dillenseger

CNET, BP 98, F-38243 Meylan cedex, france
bruno.dillenseger@cnet.francetelecom.fr

Abstract. As agent technologies are increasingly being involved in telecommunication-related applications, the need for open standards is becoming critical. During the past years, different scientific communities gave birth to different standardization actions, such as the Foundation for Physical Intelligent Agents (FIPA) and the Object Management Group's MASIF (Mobile Agent System Interoperability Facilities). Although they finally share some major targets, the OMG and FIPA current results show their distinct origins, respectively with a Distributed Artificial Intelligence and Multi-Agent Systems awareness on the one hand, and a telecommunication and information technologies background on the other hand. In a context where these two actions think about joining their achievements to upgrade each other, this article reports several experiments, carried out during the five past years in the agent platforms field, mixing both the intelligence and the middleware aspects.

1 Context

1.1 The Story

As agent technologies are increasingly being involved in telecommunication-related applications, the need for open standards is becoming critical. During the past years, different scientific communities gave birth to different standardization actions, such as the Foundation for Intelligent Physical Agents (FIPA) and the Object Management Group's MASIF (Mobile Agent System Interoperability Facility). Although they finally share some major targets, the OMG and FIPA current results show their distinct origins, respectively with a Distributed Artificial Intelligence (DAI) and Multi-Agent Systems awareness on the one hand, and a telecommunication and information technologies background on the other hand.

Reporting our work in this context is a sort of historical counterfeiting, but is a striking way of explaining and justifying a hybrid approach that we have been carrying on for more than five years. In fact, we were inspired by several needs:

- a middleware to fit the distributed applications requirements (typically in the Computer Supported Cooperative Work and groupware fields);
- a higher level layer to support knowledge representation, distributed problem solving, cooperation...

As a result, we got involved in both the middleware and the DAI communities, and we started to mix their techniques.

1.2 Multi-Agent Platforms

Many multi-agent platforms offer modelling and implementation solutions to the distribution of intelligence. To a certain extent, one may consider multi-expert systems and Distributed Artificial Intelligence as parts of the origins of the multi-agent systems. Following this point of view, we can see through the introduction, in typical centralized Artificial Intelligence languages, of low-level communication features (e.g. TCP/IP sockets in Lisp or Prolog), or more elaborate communication structures (e.g. blackboards and *Linda Interactor* in [20]), the emergence of the first multi-agent platforms. Then, sophisticated models (*actors* [1]) and techniques (constraints, reflexivity [10]) have been merged to enhance multi-agent platforms.

But, as the enhancements are going on, the resulting diversity and heterogeneity makes it difficult for a standard to emerge, besides the AI classics, which industry is just beginning to appropriate to itself. Moreover, when it comes to distribution and communication, specific solutions are often applied, sometimes through simulation. But, the more these platforms integrate to standards, in real applications, the more striking is the proof of their accuracy. Then, it seems useful to find a way of defining a standards-based bridge between the real applications and multi-agent platforms, without preventing their evolutions, but enforcing reusability and interoperability.

1.3 Mobile Agents Platforms

Whereas Sect. 1.2 was dealing with multi-agent platforms springing from the DAI community, another type of agent platform is becoming more and more popular today. These platforms deal with mobile agents, and come with a strong telecommunication and information technology background. What can be called more generally "mobile computing" addresses various issues:
- access and locally process (so-called *remote programming*) big amounts of distributed information (e.g. distributed data mining);
- the active networks field, which aims at providing networks with autonomous, intelligent and dynamic setup features;
- management and access to distributed services and electronic marketplaces;
- load balancing and fault tolerance.

First introduced by General Magic's Telescript for electronic commerce matters, mobile agents technology now benefits from the popularity of Java, which brings a precious transparency to heterogeneity. As a result, most of these platforms[1] consist in Java packages, offering an Agent class with communication and mobility features, and more or less sophisticated architectures and services. There also exists other languages-based platforms, such as Agent/TCL [8] (now called *D'agent*, as it will support other languages than just TCL in a near future), or hybrid platforms (e.g. the Tube [9], based on a Scheme interpreter written in Java).

The mobile agent community is very involved in security matters, as it is a specific and critical issue. Interoperability support between heterogeneous agent platforms, agents, and the need for intelligence are still needed, but nothing really exists at the present time. In this context, the FIPA and OMG standards are interested in playing a key-role to provide agents with interoperability and cooperation abilities.

1.4 Object-Oriented Communication Architectures

Operating systems, or layers between them and applications (*middleware*), increasingly integrate distribution and communication features. Standards from ODP (Open Distributed Processing) and OMG (Object Management Group) aim at designing a standard system environment which transparently cope with open systems issues, such as heterogeneity, interoperability, portability, distribution.

ODP brings a set of standards springing from independent organisations: ISO, IEC, ITU-T, AFNOR- The Reference Model (RM-ODP) [12] provides a conceptual framework for specifying an open distributed architecture. One key-point of RM-ODP is the use of five specification languages, bound to five points of view: enterprise, information, computation, engineering, technology. Strongly object-oriented, RM-ODP's concepts aim at improving interoperability and portability, while making distribution and heterogeneity transparent.

To a certain extent, the OMG's approach is more pragmatic, as it consists in building an open and non-proprietary object-oriented communication architecture by standardizing available technologies. Since its foundation by 11 organizations in 1989, the OMG gathered more than 500 members, among the main computing industry vendors. The CORBA standard [17] is being permanently refined and extended by the OMG, and several commercial implementations are available. The OMG's standards specify:
- an object-oriented model (*Core Object Model*);
- a reference architecture for objects management, based on the definition of the *Object Request Broker*;
- a *Common Object Request Broker Architecture* (CORBA);
- interfaces to generic objects and services: *Common Object Services* (e.g. naming service, event service), *Common Facilities*...

---

[1] IBM's *Aglets*, General Magic's *Odyssey*, IKV++'s *Grasshopper*, ObjectSpace's *Voyager*, The Open Group's *MOA*, Mitsubishi's *Concordia*, Fujitsu's *Kafka*...

1.5 Our Approach

Object-oriented distributed systems come with communication architectures supporting generic functions related to distribution. A middleware such as CORBA gives agent platforms an actual distribution facility, while abstracting from heterogeneity and relying on basic communication primitives and common services. Moreover, the integration of intelligent agents to CORBA could lead to new intelligent common services, and bring advanced solutions to generic issues of CORBA applications.

Thus, an agent platform based on a standard communication layer:
- can concentrate on its specific jobs (information finding, knowledge representation, problem solving, cooperation-);
- is likely to benefit from, as well as to enrich, the applications and services of the environment it integrates to;
- builds the *bridge* we deal with in Sect. 1.2.

Today, some on-going work in the *agent community* and the OMG tends to encourage such a bridge. FIPA, for instance, underlines that "*agents need to be able to integrate with, and where appropriate use for themselves, existing and emerging computational infrastructure. Examples include: TCP/IP networking, CORBA, TINA-C, http and OLE.*" [4]. As far as the OMG is concerned, the MASIF [16] include a set of definitions and interface specifications related to agents systems interoperability. An agent is described as "*a computer program that acts autonomously on behalf of a person or organization. Most agents today are programmed in an interpreted language (for example, Tcl and Java) for portability. Each agent has its own thread of execution so that it can perform tasks on its own initiative*". As we'll see in this article, these two last sentences have also something to do with our work (cf. "interpreted language" and "thread of execution").

2 The First Experiments

2.1 PUMA, the Forerunner

PUMA (Prolog Upgrade for Multi-Agent world) is our first experiment [5]. It results from the integration of a Prolog interpreter in a C++ object of the COOLv1 object-oriented communication layer [14]. The Prolog kernel and dialect have been extended to integrate COOL-based communication primitives: message and mailbox based communication, including group features, local synchronous communication, naming service, mobility, COOL object creation. Each agent has its own thread of activity, and can explicitly move and transparently resume its activity at the next programme goal.

PUMA have been enriched with many C++ development classes and Prolog modules allowing the incremental and modular composition of the agent's behaviour:

activity step, cooperation structures and protocols, knowledge representation, specialised services... The system has been used to implement a distributed multi-agent application for meeting-rooms booking, accordingly to a multi-agent approach for office information systems [6]. Cooperation was based on communication groups with specific management and invocation protocols, dynamically created mobile agents for sub-contracted tasks, and constraints-based negotiation.

The dynamic, modular and declarative programming of agents, the activity-safe mobility, the constraints-based negotiation language, the group organization for cooperation, but also the quick prototyping opportunity, are the key-points brought to light by the PUMA experiment. Its career was interrupted by a big evolution of COOL, which evolved towards a CORBA conformance. However, other experiments were necessary, not only to follow this evolution, but also to try other combinations with other interpreted languages. Another practical aspect about the interpreter kernel is that the integration we did in PUMA needed much development time on the Prolog kernel source code, not only to add new primitives, but also to encapsulate it (break the interpreter loop and build a control function, extract and encapsulate the global variables to create several independent Prolog objects in a multi-threaded process).

2.2 The Successors

The immediate successors to PUMA followed the evolutions of COOL. This layer gained more and more distance from the underlying micro-kernel specificities (Chorus), and progressively became CORBA compliant. COOLv2 figures an intermediate state, mixing typical COOL features (persistence, mobility, communication groups, activity and concurrency management) with a CORBA-like architecture. COOL-ORB [3], the latest evolution, comes with a fully CORBA compliant platform. Beyond the middleware concerns, we also tried different kinds of integration of other logics-based interpreted languages, featuring extensions such as constraints.

Compared to PUMA, ROZACE (Remote execution server for OZ Agent in CORBA Environment) figures a complementary approach [15], by keeping the interpreter kernel apart from the agent and making it available as a script execution server. This architecture illustrates the case of a really big interpreter kernel, which offers many powerful features, but needs a large amount of computing resources. As a result, agents are *light*, but can perform advanced tasks by submitting scripts to the remote execution server[2], either synchronously or asynchronously. ROZACE was built on COOLv2 and Oz [19], an interpreted language that combines the logic, constraints, concurrent, high-order functional and object-oriented programming paradigms. Unlike PUMA, no primitive has been added to the language, and scripts are pure Oz. The Oz constraints features have been applied to a few parts of the room booking application formerly developed with PUMA.

CHOCOLAT (CHorus COol & Life Agents Tool) [11] has been developed in parallel with ROZACE, to explore a PUMA-like integration on COOLv2 (i.e.

---

[2] In fact, this approach can be mixed with PUMA's: a small interpreter kernel may be integrated in the agent to implement a minimal autonomous behaviour and knowledge representation.

integrated interpreter, extended dialect), but relying on an extended logic programming language. Derived from Prolog, LIFE (Logic, Inheritance, Functions and Equations) offers functional programming, concurrency and constraints features, with types (cf. *sorted logics*) and inheritance [2]. The quality of LIFE's C language interface allowed a quick integration, with fewer source transformation than in the case of PUMA. Thanks to the underlying COOLv2 middleware, it has been possible to introduce a dynamic invocation mechanism, allowing the CHOCOLAT's interpreted kernel to invoke any method on any COOLv2 object, with no preliminary link edition. CHOCOLAT has been used to implement a representation and revision model for the beliefs of agents [18].

2.3 Conclusion

These experiments taught us some tips about middleware-based agent platforms. Not wanting to start from scratch, most of the work often consisted in adapting high-level interpreted languages from the source code, which is not convenient, neither efficient. In fact, we realized that very few such languages are designed as ready-to-integrate components.

As far as the agent primitives are concerned, those dealing with message sending, mailbox management and address publication (cf. naming service) appear to be essential and straightforward to implement. Group features, including broadcasting and functional sending, are of great interest for agents organization and cooperation matters. The dynamic creation of agents is also very useful, as it makes it possible for an agent to sub-contract a task to an autonomous extra activity.

Synchronous communication, which typically consists in having another agent executing a script in a synchronous way, generally causes concurrency problems at the embedded interpreter kernel level. As a result, this kernel has to be protected by mutual exclusion, which causes a deadlock in case of cyclic synchronous calls. So, this can be a convenient feature, but it must be used very carefully.

The main difficulty springs from mobility. PUMA was the only platform supporting a transparent agent state and activity mobility feature. It relied on COOLv1's migration feature and on Prolog's save/1 predicate, which creates a complete state file. But this mobility was limited to homogeneous environments.

3 Building a Multi-Agent Application on CORBA

3.1 What is "CIDRIA Générique"?

CIDRIA Générique is a generic workflow system, built on CORBA and intranet technologies, according to a multi-agent model [7]. *Workflow* systems consist in executing predefined or on-line generated scenarios, in order to trigger and synchronize a set of tasks in the information system, while achieving information gathering, circulation and tracing. CIDRIA Générique aims at managing procedures combining any kind of service, involving any type of resource: software, hardware, users... To cope with this real world heterogeneity, a homogeneous multi-agent virtual world is created by mapping each resource to a dedicated agent. Inside the multi-agent world, resources are represented by their *skills*, and they *cooperate* via *service requests* accordingly to these skills.

3.2 The Multi-Agent Platform

While making CIDRIA Générique, we wanted to produce as reusable as possible developments. We chose an on-the-shelf CORBA compliant middleware, and a Edinburgh-type Prolog interpreter (SWI-Prolog [21]). The high quality of its C interface made it possible for us to embed it in a C++ class without modifying the source code. This way, any interpreter or middleware update has a minor impact on our work. This class has been used to integrate a Prolog kernel in a CORBA server. As a result, we could implement every server operation in Prolog, through the C++ mapping[3]. This integration is minimal, as we didn't extend the Prolog dialect with communication primitives. In current version, agents are concentrated on the same server, and share, for practical reasons (e.g. persistence, resources mobility transparency), the same Prolog kernel. Nevertheless, agents cooperate and communicate via message handling (no shared memory is used), and the architecture is ready to be extended with actual agents distribution, with several servers.

3.3 Conclusion

Built following modularity and reusability principles, CIDRIA Générique fits evolution. When developing and maintaining it, we really appreciated the comfort of Prolog programming, propitious to quick prototyping. As the agents are fully represented in Prolog, saving their states as a set of Prolog clauses fully describes the overall system. We used this opportunity for debugging, and also to introduce persistence features. Moreover, the Prolog declarative programming makes it possible to dynamically modify the operations implementation, without interrupting the server functioning. One just have to modify a Prolog file, and "reconsult" it from the server; the agents' behaviour is updated while their states are not affected (unless desired).

From a performance point of view, one could believe that the CORBA+Prolog association is too heavy, but it doesn't appeared to be the case. Although we didn't perform any load test, demonstrations made through the french network RENATER, between Paris (INRIA) and Caen (CNET), didn't reveal any particular response delay when invoking the server's operations; i.e. it looked like a local application[4].

---

[3] A CORBA server interface is declared independently from the implementation language, in IDL (Interface Definition Language). Applying this abstract definition to a given programming language, to make server or client software, needs a specific *mapping*.

[4] The client machine was a portable PC running Windows 95, and the server machine was a Sun UltraSparc 1 with 64 Mo of RAM, and wasn't dedicated to the demonstration (it run

To conclude, if our previous experiments tend to prove that object-oriented middlewares can bring an accurate solution to the communication needs of agent platforms, CIDRIA Générique legitimates the use of a CORBA-based multi-agent approach for the conception and implementation of distributed applications.

4 Recent Work

4.1 An "Interpreter" CORBA Interface

In order to extract the generic aspects of developing multi-agent platforms by integrating interpreted languages to CORBA, we have defined an "interpreter" CORBA interface, named `Interp`. It specifies two operations:

- `execute(in string script, out string result)` makes the server object execute `script` and give the `result` of the execution;
- `receive(in string message)` makes the server object add `message` to its mailbox. No message format is specified. It may consist of a script to run asynchronously, or any kind of data.

Messages, scripts and results are generically represented by strings. If a problem occurs during a call, these operations raise an exception. There are 4 exceptions:

- `FULL_MAILBOX` indicates that the message couldn't be added to the mailbox;
- `SCRIPT_ERROR` means that the script couldn't be properly executed, because of a syntax or execution error;
- `NOT_IMPLEMENTED` is raised when the operation isn't available in the server object;
- `_UNKNOWN` informs that the operation couldn't be performed, but provides no diagnostic.

4.2 A Generic ORB Class

As we were to extend an interpreted dialect with communication primitives based on the middleware, accordingly to the `Interp` interface, we decided to generically (i.e. independently from the actual interpreted language) encapsulate these primitives in a C++ class, named `Generic`. We call it `Generic` because it isolates the interpreter kernel from the actual middleware. But this class is middleware dependent, and needs to be adapted from one middleware to another. Nevertheless, adaptation is clearly straightforward between CORBA compliant platforms[5].

Public methods of the `Generic` class include:

---

other server processes). Client software is written in Java, and server software in C++ and Prolog.

[5] Group communication primitives are specific to the ORB we chose (COOL-ORB), and should be implemented on other CORBA platforms via specific servers.

- initialization and destruction methods, which manage, among others, the link with the underlying ORB (e.g. access to the naming service, the group service...);
- two *pure virtual* methods[6], `run()` and `call()`, respectively representing an interactive interpreter loop, and a script execution method. These methods will have to be implemented when deriving the class for a given interpreted language;
- generic communication primitives, which are likely to extend the embedded interpreted language: naming service, message sending, mailbox management, communication groups, synchronous remote script execution, object (agent) creation;
- two methods implementing the `Interp` operations (see Sect. 4.1).

4.3   ORB-fying an Interpreter Component

Theory. Suppose we find a C++ class interpreter component, that we can derive, and that makes it possible to create several objects in the same multi-threaded process (typically one object and one thread per agent). This class also provides:

- a way to add primitives, by adding or overriding one or several methods;
- facilities to handle the entities of the embedded language (e.g. Prolog terms, or Lisp expressions);
- methods for external interpreter loop control and script execution, with error handling;
- conversion methods between the string and internal representations of the language entities, which really helps write the message sending and script execution primitives.

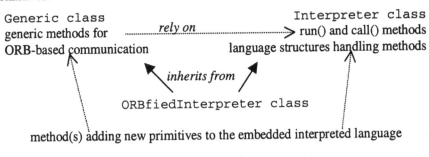

Fig. 1. The ORBfiedInterpreter class defines new primitives based on the inherited ORB-based communication methods (Generic class) and interpreted language structures handling methods (Interpreter class). The Interpreter class must define two specific methods to complete the Generic class' implementation (run() and call()).

Provided the fact that we find this ideal component, the integration task is easy to do through multiple inheritance, as shown by Fig. 1, and we get a CORBA-based agent

---

[6] A *pure virtual* method consists in a method declaration without implementation. Such a method can be called, but the class it belongs to is *abstract*, i.e. it cannot be instantiated. This class has to be derived so as to actually define the method, and thus be able to create objects.

platform. The port to another communication layer just involves the Generic class, and the port to another operating system is really minimal, while achieving interoperability.

Practice. Confrontation to reality rapidly destroys this dream. According to our investigations, "AMZI! Prolog + Logic Server" is the only available interpreter component which matches our main requirements. Most of the others come with poor C interfaces which really turns the integration task into a nightmare. Obviously, global and static variables forbid multi-threading and multiple independent interpreter kernels in the same process. The worst C interfaces keep the main activity in the interpreter loop, which makes it very hard to implement a synchronous call primitive. Sometimes, the string conversion functions for the language entities aren't directly available, whereas they always exist in some form in any interpreter.

Finally, we decided to build by ourselves the libraries and their C++ encapsulation from existing interpreter C sources. We believe that on-the-shelf C++ components will be soon available, following AMZI! Prolog, Ilog products, or a few *extension languages*[7] way. Then, we'll just have to change our adapted kernels for these ready-to-integrate components.

4.4 A Prolog Integration: BPorb

BinProlog is an Edinburgh-style Prolog implementation [20]. It is famous for its high performance, and the C source code is available at a low price academic licence. Its C++ encapsulation has been very tricky, for its C interface really didn't fit our needs. The result is interesting, however, as we succeeded in overcoming the fact that BinProlog keeps the main activity in its own interpreter loop. This was achieved thanks to a BinProlog primitive which creates a new engine to solve a goal, without destroying the current goal.

The resulting `BinProlog` class mainly supplies the `call()` and `run()` methods, and an `extension()` virtual method. This method is invoked by the Prolog kernel to implement the new_builtin/3 predicate. The new_builtin(+code, +in, ?out) goal is mapped to a call to `extension(code, in, out)`, where code figures a primitive number. A null return value from `extension()` means a goal failure.

Once this encapsulation is done, the final step consists in defining the `BPorb` class, which inherits from `Generic` and `BinProlog`. The `extension()` method is overriden, to map the communication primitives provided by `Generic` to the Prolog dialect, via the new_builtin/3 predicate.

---

[7] The extension language principle consists in integrating an interpreted language into an application in order to provide users with a powerful configuration and customization tool. Choosing a standard language prevents users from learning a new language for each application. The Emacs Lisp and the Guile extension language, from the Free Software Foundation, illustrate this idea. Elk (see Sect. 4.5) is another example of extension language which shows Scheme's popularity.

4.5   A Scheme Integration: ElkOrb

Elk [13] is a free Scheme implementation, specially designed to be embedded (see *extension language*, note 7). Its C++ encapsulation resulted in two classes:
- the `ElkObject` class encapsulates the Scheme entities, with handling methods for every type;
- the `Elk` class encapsulates the language engine.

The `Elk` class mainly defines the `call()` and `run()` methods, required by the use of the `Generic` class. `Elk` also defines an `extension()` method, which is called when evaluating the (new-primitive arg1 arg2 ...) expression. The `extension()` method takes a vector of `ElkObject` objects (arg1, arg2 ...) as an argument, and returns an `ElkObject` object, which represents the evaluation result.

Once this encapsulation is done, the `ElkOrb` class is obtained by inheriting from `Generic` and `Elk`. `ElkOrb` overrides the `extension()` method to make the Generic communication primitives reachable from the Elk kernel.

4.6   Observations

The `ElkOrb` and `BPorb` classes have been used to create slave or interactive interpreter objects (slaves permanently look for incoming messages and execute them as scripts). At initialization time, these objects load a Scheme or Prolog file which contains their behaviours. They consist in CORBA server objects, registered in the naming service under chosen names which represent their addresses for messages and synchronous calls. Each communication group is also registered in the naming service, but under two names: one for message broadcasting, the other for *functional* (i.e. an arbitrary member of the group receives the message) synchronous or asynchronous calls.

Since communication primitives are based on the `Interp` CORBA interface, it is possible for different interpreted languages, integrated to CORBA via the `Generic` class, to communicate asynchronously or synchronously with each other. The main issue is to build a script for another language, and then to extract the result, via a string representation. It could be interesting to look at Xerox's ILU approach. The Inter-Language Unification platform formerly aimed at inter-language interoperability, but then took distribution into account, and finally offers a CORBA hook.

5   Conclusion

The experiments we have reported aim at proposing a bridge between distributed applications and multi-agent platforms, based on the adoption of a standard and non-proprietary middleware, CORBA. The reasons for such an approach are:
- CORBA platforms, services and distributed applications can be enhanced with intelligent features;

- Distributed Artificial Intelligence techniques may find a concrete investigation and application field;
- agent platforms may abstract themselves from the low-level distribution management primitives (basic communication, CORBA services), while adopting a standard communication architecture, propitious to their interoperability.

We have integrated typical Artificial Intelligence languages to CORBA platforms, either as remote execution servers, or by embedding an interpreter kernel in the objects, and extending its dialect with CORBA-based communication primitives. The last experiment resulted in extracting some generic features, thus isolating the interpreter part from the CORBA communicating part.

With the recent outcome of FIPA's and OMG's actions, and the will to combine the former's ACL (Agent Communication Language) with the latter's MASIF specifications, our experiments could be regarded as a way of combining a high-level approach with a middleware support.

Our further work will focus on agent mobility, OMG's and FIPA's work, CORBA services and facilities, and CIDRIA Générique's evolution (integration of our last results, extensions with actual distribution and agent communication protocols). As a matter of fact, this application will be our privileged integration and demonstration platform for truly standards-based mobile intelligent agents.

Acknowledgements

The work reported in this paper was carried out in the CNET[8] lab in Caen, by Christophe Trompette, Emmanuel Hym, Eric Malville, Stéphane Piolain, and Bruno Dillenseger, under the responsibility of François Bourdon and Anne Lille, in collaboration with Patrice Enjalbert and Jurek Karczmarczuk from the University of Caen.

6 References

1. Gul Agha. Actors, a Model of Concurrent Computation in Distributed Systems. MIT Press (1988).
2. Hassan Aït-Kaci, Bruno Dumant, Richard Meyer, Andreas Podelski, Peter Van Roy. *The Wild LIFE Handbook (prepublication edition)*. Digital Paris Research Laboratory (1994).
3. *Chorus / COOL-ORB Programmer's Guide*. Documentation CS/TR-96-2.2, Chorus Systems (1997).
4. Ian Dickinson. *nyrequirements*. FIPA document notes, Yorktown (September 28th 1996).
5. Bruno Dillenseger, François Bourdon. *Supporting Intelligent Agents in a Distributed Environment: a COOL-based Approach*. TOOLS 16, Prentice Hall (1995) 235-246.
6. Bruno Dillenseger. *Une approche multi-agents des systèmes de bureautique communicante*. Thèse de Doctorat, Université de Caen (1996).

---

[8] "Centre National d'Etudes des Télécommunications", France Télécom research laboratories.

7. Bruno Dillenseger, François Bourdon. *Modélisation de la coopération et de la synchronisation dans les systèmes d'information (une expérience de workflow basée sur les nouvelles technologies).* Calculateurs Parallèles Vol. 9, No 2, HERMES (1997) 183-207.

8. Robert Gray, David Kotz, Saurab Nog, Daniela Rus, and George Cybenko. *Mobile agents: The next generation in distributed computing.* In Proceedings of the Second Aizu International Symposium on Parallel Algorithms/Architectures Synthesis (pAs `97), Fukushima, Japan. IEEE Computer Society Press (1997) 8-24.

9. Dave Halls. *Applying Mobile Code to Distributed Systems.* Doctoral Dissertation Computer Laboratory University of Cambridge (1997).

10. Salima Hassas. *GMAL - Un modèle d'acteurs réflexif pour la conception de systèmes d'intelligence artificielle distribuée.* Thèse de doctorat, Université Claude Bernard - Lyon I (1993).

11. Emmanuel Hym. *Intégration et mise en oeuvre d'un langage interprété sur la plate-forme répartie à objets Chorus/COOLv2.* Rapport de DEA, Université de Caen (1995).

12. ISO, ITU. *Reference Model of Open Distributed Processing.* ISO standard 10746, ITU-T recommandations X.900 (1995).

13. Oliver Laumann, Carsten Bormann. *Elk: the Extension Language Kit.* Technische Universität Berlin / Universität Bremen, Germany (1996).

14. Roger Lea, Christian Jacquemot, Eric Pillevesse. *COOL: System Support for Distributed Programming.* Communications of the ACM, Vol. 36, No 9 (1993).

15. Eric Malville. *Etude d'une architecture agent sur la plate-forme Chorus/COOLv2.* Rapport de DEA, Université de Caen (1995).

16. *Mobile Agent System Interoperability Facilities Specification.* Joint submision: GMD Fokus & IBM Corp., supported by Crystaliz Inc., General Magic Inc., The Open Group. OMG TC document orbos/97-10-05 (1997).

17. Object Management Group. *The common Object Request Broker: Architecture and Specification (revision 2.0).* OMG (1995).

18. Stéphane Piolain. *Expérimentation de l'approche Programmation Logique Etendue pour la réalisation d'agents nomades sur Systèmes Répartis à Objets.* Rapport de stage de DEA, Université de Caen (1996).

19. Gert Smolka. *An Oz Primer.* DFKI Oz documentation series, German Research Center for Artificial Intelligence (DFKI), Stuhlsatzenhausweg 3, D-66123 Saarbrücken, Germany (1994).

20. Paul Tarau. *Bin Prolog 5.75 User Guide.* Département d'Informatique, Université de Moncton, Canada (1997).

21. Jan Wielemaker. *SWI-Prolog 2.5.* Dept. of Social Science Informatics (SWI), University of Amsterdam, Roeterstraat 15, 1018 WB Amsterdam, The Netherlands (1996).

Open Scalable Agent Architecture for Telecommunication Applications

Dirk Wieczorek; Dr. Sahin Albayrak

Technical University of Berlin
DAI-Laboratory
Email: {wieczore|sahin}@cs.tu-berlin.de

Abstract. This paper describes the JIAC project, an open and scalable agent architecture. The telecommunication market of today being in continuous expansion is in need of a platform to access the resulting growing demands. JIAC-Java Intelligent Agent Componentware-provides an agent based platform to meet those demands of the telecommunication applications. The requirements of those applications are outlined in detail in the 3rd chapter. Agent oriented techniques, as described in the first sections, are considered the adequate solution for the maintenance of networks and the provisioning of services. JIAC is using a component based approach to build customized agents. To realize telecommunication applications JIAC relies on the metaphor of electronic marketplaces as the basic structure.

1   Introduction

The vision of the telecommunication market today is *"information at any time, at any place, in any form"*. Within this open market of information services, the aspects of service customisation and instant service provision are of fundamental importance. Moreover, the provision of telecommunication services will be from differing network platforms to a large variety of terminals (e.g., telephone, portable computers or personal digital assistants (PDAs)). In this context, a new paradigm, namely the paradigm of agent oriented techniques (AOT), is gaining momentum. The main actors of this scenario are autonomous, mobile agents, which are most suitable to address all the problems related to the open market in general.

In consequence many areas have started to develop agent based solutions for telecommunication applications, intelligent networks and information retrieval. Also the development is aiming at the whole area of software engineering with agents.

Although the main topic of AOT are open decentralized systems, the basic notion of this domain is the notion of an agent. An agent is a piece of software which acts on behalf of a user or assists a user . More specifically, an agent is a self-contained software element that is responsible for performing a part of a programmatic process.

Often the specialized term "Intelligent Agent" is used, which is associated with various expectations and is used in many different contexts. Such "Intelligent Agents" often have different problem solving mechanisms, ranging from pre-defined rules up to self-learning AI inference machines. Intelligence is mostly attributed to the latter ones. However, in this paper 'intelligence' is seen more as an emergent property of certain agent systems which seem to act 'intelligently' in the eye of an observer. That means no definition of intelligence is proposed here but the notion introduced by [brooks] is used. However, it is sometimes useful to measure the intelligence of an agent system. That could be done by examining the behaviour of the system with regard to the existence and peculiarity of the attributes reflexivity, adaptivity, cooperation and autonomy of the agent system in question.

Agent technology is considered in a wide range of telecommunication applications, such as user interfaces, mobile computing, information retrieval and filtering, smart messaging, smart home, multi-media on demand, network management and the electronic marketplace for telematic service provisioning. Also it has to be recognised, that agent functionality differs considerably within these different applications. To get the most benefits from agent oriented techniques it is necessary to develop a comprehensive definition of agenthood and to find a paradigm for software development with agents. The next section will give an impression on how to do this.

1.1   Using Agents

The following attributes are considered fundamental for a piece of software to be called an agent: **autonomy, interactivity** and **reactivity.**

Since the above attributes apply to many kinds of software, it is necessary to introduce the following additional attributes in order to distinguish agent oriented software from other pieces of software, **goal-orientation, mobility, adaptivity, planning capabilities** and **cooperation.**

For the deployment of agent technology one has to develop a methodology for software engineering based on the fulfilment of the aforementioned attributes. Agent oriented software development (AOS) is such a new methodology to tackle the problem by extending object oriented design methods in some way.

The object oriented design methods comprise the construction of objects or classes, which meet the requirements of the system, a description of the object behaviour and a description of the communication relations between these objects. One can distinguish three main phases with several sub processes in the object oriented design.

The first phase is the determination phase in which the classes, the attributes and finally the collaboration are determined. The class hierarchies, the crucial parts of the system and the protocols are defined in the second phase (analyse phase). The last phase, called the protocol phase, is done in parallel to the other two phases and collects the results of them.

A preliminary system design is finished just after the analyse phase. That design is based on the derived classes from an interpretation of the system requirements. These classes are described in terms of their methods, their data and the communication relations at specific system states. The verifications, changes and optimizations of these classes are done in the third phase.

The process of AOS, following the object oriented design methods, can also consists of three parts which are **Agent Oriented System-analysis**, **Agent Oriented Design** and **Agent Oriented Programming**.

The first phase, agent oriented system analysis, is concerned with the identification of the agents which exhibit the desired system behaviour. Each agent has its own unique set of capabilities which characterise it. In the second phase, agent oriented design, the interactions (agent language, protocols) between these agents and their roles are defined. The roles are a metaphor to describe the default behaviour of an agent in the system, such as trader and seller.

Another aspect of agents in a multi-agent system is that they are usually organized in a hierarchy. That is resembled by the organization model which is also defined in the design step. And last but not least a suitable architecture is chosen and the agents are implemented according to the specifications of the preceding two steps.

The benefits of such a methodology are fast development cycles if relying on existing agent architectures and high reusability because the system behaviour is mainly defined by the capabilities of an agent.

1.2 Agents for Telecommunication Systems and Applications

The telecommunication market is expected to be one of the most crucial markets of the future. It is already one of the most rapidly developing areas, especially in terms of deregulation and strategical impact.

As a consequence network providers are forced to develop from network providers to service providers that offer the customer not only simple network use, but complete value added services within the network, i.e. the internet. A secure, reliable and integrated approach to provide services and operation is required. Therefore a generic and secure platform like Java, enhanced by several service management capabilities and agent technology is privileged for that purpose. The agent oriented approach to fulfil these criterias seems to be one of the most promising ones. Only with autonomous, dynamic entities (agents) it is possible to adapt to the differences and requirements in the heterogeneous networks of different providers during runtime.

To meet those demands, agents for telecommunication applications should have a set of features for this domain. Those features which are directly deduced from the demands should include:

- **Scalability** means, that an agent system can be adjusted to the needs of systems of various complexity. The scalability feature of an agent system will address the problem of the scalability of Intelligent Networks.
- **Manageability** means, that those agent systems should have a fault management to carry on even if parts of it cease to work. Furthermore, in order to adapt to different users or environments the agent systems must be configurable to an extend.
- **Security** is an important issue for e-commerce applications and electronic marketplaces. That's why agents should support secure communication and protocols.
- **Accounting** (the process of collecting data about transaction) and **Billing** (making an invoice based on the accounting data) for e-commerce should be included.
- Support of net **standard protocols** and platforms like TINA[dpe] must be included to integrate the agent system in existing platforms.
- The agent system should offer means to **wrap existing legacy systems** to offer a uniform method of access that will allow an easy integration of databases from different service providers.
- **Personalizability** means, to be able to customize an agent such that it carries out tasks that are useful for its user, which may or may not resemble tasks carried out by other agents for other users. This customization could come from explicit programming, or from machine learning. In the latter the agent may infer user behaviour from observation.
- **Openness,** that allows the use of existing applications and easy upgrading.
- The integration of **user interfaces** should be supported to shorten development time and to increase user acceptance.

If agents are used for telecommunication applications, especially for service provisioning in connection with mobile clients, it is necessary to have a dedicated infrastructure for the agents to live in. The structure is deduced from the metaphor of electronic marketplaces.

1.2.1 Electronic Marketplaces a Platform for Agents

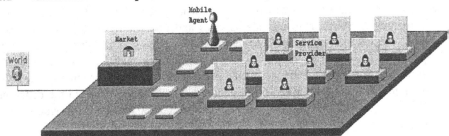

Fig. 1. An electronic marketplace

An **electronic marketplace** as shown as an example in figure 1 is defined as an open and distributed platform where service provider can offer their services, represented by service provider agents. These marketplaces provide an interface to clients and vice versa to the provider to make transactions in the sense of communication, money / product exchange. **Service provider agents** offer services and a suitable interface to use these service. **Content provider agents** allow access for data that is needed by the service provider. It may be viewed as a special service provider for connecting to a legacy system. The services on the marketplace are usually accessed by **mobile agents,** which rely on secure migration protocols offered by the markets.

A mobile agent can be observed on the figure 1. These agents travel along several marketplaces in order to find a suitable service provider agent. **Mobility** of agents is one of the features which are supported through the use of marketplaces. This allows agents to move around and enables them to perform their tasks locally, at the locations where the involved resources/entities are located, rather than "shouting" requests across the network in order to access resources remotely. Consequently, the underlying concept is often referred to as "Remote Programming" (RP). In connection with electronic marketplaces the feature of mobility of agents might help to reduce online cost by offline access. The advantage is that a mobile client can invoke an agent request during a brief connection to the network. Later the client is able to fetch the agent and its containing information on a second subsequent session. These type of connectivity allows the client to have a relatively low-bandwidth connection like a mobile telephone. Because the information is reduced by the agent in a retrieval and filtering process on the server itself and therefore the actual data transmitted is minimized. Also the clients machine, having the majority of computing done on the server, need not to have large processing capabilities. In effort to protected the user from data overload and to allow efficient information retrieval a peronalized, mobile agent being able to represent the users desires to search engines will really make a difference. So the agent is able to conduct real semantic searches according to the users preferences within the network.

The difference between mobile code or agents and message is that a user can send 'intelligence' to the destination to act according to the wishes of the user. That is to send information and the means to handle it rather than sending information only.

Other features available through the use of electronic marketplaces include secure communication, certification management, embedded running of mobile agents and management of service providers. The remainder of this paper is concerned with a framework called JIAC which will help to develop agent based applications relying on electronic marketplaces.

1.3   JIAC - The Toolkit for Agent Oriented Telecommunication Applications

JIAC, the abbreviation of Java Intelligent Agent Componentware, is a framework that addresses the three parts of the before mentioned agent oriented software development process.

With the ongoing success of the Java programming language and its suitability for mobile and platform-independent applications one can see an upcoming popularity of

Java in the mobile agents area. Although some interesting systems of mobile agent environments for Java are already available, JIAC realizes some unique features for the realization of telematic applications like support for electronic commerce, highly scalable architecture, wrapping of legacy systems with exchangeable I/O components.

JIAC is an implemented Java API that offers component based mobile agents. Beside the aspect of distributed, mobile agents, JIAC features a complete set of structures, which are realized by the notion of marketplaces, like accounting, billing and security for electronic commerce applications or distributed provisioning of telecommunication services.

The process of software development is supported by tools like an agent oriented debugger. In the following JIAC is described especially on the aspects of componentware and support for telecommunication applications.

1.3.1 Agents and Marketplaces

In, JIAC agents are entities that have reactive and asynchronous task processing. The agents pursue their goals autonomously that involves the migration to marketplaces to use offered services the agent needs to fulfil its task. However, the user of an agent can get notified by the agent from the remote location that the agent migrated autonomously. By using certain tools like an "agent navigator", the user stays in control of the agent.

In order to meet the demands of applications JIAC has basically two types of agents:

- **static agents** which exist in marketplaces and offer services. They are fixed to a certain marketplace. The following types of agents may be considered static agents:
  - manager agents, which administrate a marketplaces, controlling the migration, provide an embedded running of mobile agents and act as a message distribution centre.
  - service provider agents, offering value added services.
  - content provider agents, providing uniform access to legacy systems.
  - user interface agents, being the bridge to the user in the JIAC system.
- **mobile agents** which can migrate from marketplace to marketplace.

Furthermore, the agents provide a dynamic, on-demand combination of several features and services, by cooperation and interaction with other service providing agents -- but transparently for the user.

Following we focus on the counterpart of our agents, the *marketplaces*.

In simple terms, *JIAC-Marketplaces* are Java software platforms, where several agents can gather to offer several services. The marketplaces are organized in certain hierarchies and administrated by special *manager-agents*. These manager agents offer an interface for the use of the services, that the agents of the marketplace provide. Another aspect that is covered by the manager agent is to provide the facilities for migration of mobile agents. Furthermore, it takes care of infrastructural topics on a marketplace like *logging* and *security*.

Those marketplaces are not limited to one host. Also, it is quite easy to distribute a marketplace via several host and, due to Java , via several platforms. Though, a manager agent is associated with every marketplace, there are a number of approaches to manage marketplaces and resident agents on them. One of those approaches is to used address files. An address file contains an address table of agent names and their locations just like a yellow page. But static address files are neither used by the manager agent of a marketplace nor by agents, since they have a number of drawbacks.

These drawbacks are addressed by a different, more flexible approach which uses a default port and a special marketplace structure.

1.3.2 Communication

In JIAC all agents communicate with each other by the use of speech acts. The protocol for the speech acts follows the proposed KQML (Knowledge Query and Manipulation Language) standard.

A speech act as it is used in JIAC is defined as a tuple consisting of

- Sender address of sending agent,
- Receiver address of receiving agent,
- Content subclass of a certain knowledge class,
- Ontology the context of the content,
- Reply_With, a tag which correlates questions and answers and
- In_Reply_To the reverse of the above.

Apart from the mentioned standard KQML attributes these speech acts feature some extensions, which allow the realization of secure and reliable migration protocols. In JIAC it is taken care that a mobile agent is held frozen on its start market until its transmission to destination market was successful.

JIAC supports the following speech act types: **Tell, Untell, AskOne, Sorry, Reply, Evaluate, Register** and **Unregister**.

1.3.3 Scripts

Scripts, the capabilities of an agent, are an important key concept in the architecture of JIAC. They are a convenient way to specify the agents behaviour. Those scripts from the set of possible actions the agent can perform, defining the agents place within a particular application. By holding more than one script, the agent can change its behaviour dynamically, by just activating an other script of its stock, given a certain stimulus. This makes it possible to personalize and customize the agents in a very fine granularity, which is essential for heterogeneous environments like telecommunication networks.

As it is shown in figure 2 the script consist of a number of script elements. These elements are necessary to allow the agent to migrate while keeping his current state. Since the serialization, which is used to migrate agents, does not permit to serialize threads the states of a script are maintained through the use of these script elements representing some kind of state machine.

```
<Script> :: <initScriptElement>
               {<ScriptElement>}*


<ScriptElement> ::= <ScriptElementName>:
                    <BLOCK>
                                              RETURN
<ScriptElementName>


<BLOCK>                  ::= <JavaCode>
<initScriptElement> ::= <ScriptElement>
```

Fig. 2. The structure of a script

1.3.4 Services

Services are used in JIAC to define the public capabilities of an agent. Each service is associated with a certain script. Also there are two types of services free services, being free of charge, and commercial services requiring to be paid for.

The notion of a service is defined as:

Service being a 4-tuple (S, P, E, K), whereas

- S denotes the trigger speech act
- P denotes a flag indicating whether this service is commercial or not
- E denotes a flag indicating whether this service can be used by other agents or not
- K is a flag allowing multicast of this service

To address those services comprehensively, the services can be collected in groups called categories. These categories are defined as a tuple consisting of (Name, Options, SuperCategory). Where the *Name* is the category definition, *Options* include a human readable description of the category and *SuperCategory* allows a hierarchic structuring of categories.

To address services over the boundaries of one marketplace the following approach is used.

1.3.5   Hierarchies of marketplaces

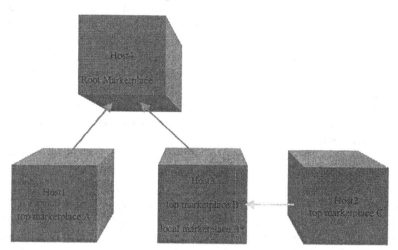

Fig. 3. A hierarchy of marketplaces

Basically, three types of marketplaces can be distinguished, a root marketplace, a top marketplace and a local marketplace which are described in the following.

Let T be a set of top marketplaces and L be a set of local marketplaces, Name, Name* symbols for unique identifiers. Finally A denotes a set of Agents.

A top marketplaces T is defined as the 4-tuple (Name, Name*, S, $A*$), where Name is the name of the marketplace, Name* the name of the superior marketplace, S a set consisting of two subsets $T*$ of T and $L*$ of L and $A*$ a subset of A. If T does not have a superior marketplace it is called root marketplace. Every top marketplace occupies the same port as all other top marketplaces on its particular host.

A local marketplace L is defined as the 3-tuple (Name, Name*, $A*$), where Name is the name of the marketplace, Name* the name of the superior marketplace and $A*$ a subset of A. In contrary to top managers it is required for local marketplaces to have a superior marketplace. So, as a result, local marketplaces require a top marketplace running on its host.

The root marketplace acts as a central instance to connect all underlying marketplaces. With the existence of such a marketplace it is possible to broker services over the complete hierarchy of marketplaces. Although such a root marketplace is not necessary, it is recommended. Without it, it is not possible to find or offer services beyond the local boundaries of marketplaces being then unsorted sets.

Now let us consider figure 3. Underneath the root marketplace on host 4 in this marketplace hierarchy are two top marketplaces A and B. Following down the hierarchy, one can find the local marketplace B* on the same host 3 as B. B* is then a sub marketplace of B. And finally, the top marketplace C being a logical sub manager of top marketplace B, can be identified on the right hand side of the picture.

The residents on these marketplaces are stationary and mobile agents. A special stationary agent is the manager agent taking care of the marketplace.

1.3.6 Brokering - in the Search of Services

The search for a service provider agent starts at the service requesting agent and moves on into the direction of the root manager. At the start of the searche, the service requesting agent looks through his own components to find a service provider. If that does not succeed the request is given to the superior agent. This agent searches his own components and, in the case of a manager agent, asks all registered agents. If no suitable agent can be found the request is given once more to the superior agent. This process moves on until no superior can be found anymore or a service provider is finally found. In the first case a failure message is generated and in the second case the service is requested.

It may happen that the whole hierarchy of manager agents is being searched through. Since that is a very costly process, the data may be cached to allow efficient retrieval of service providers.

The process of searching for a service provider can be pruned earlier, if the knowledge of the manager agents about the categories of services is heeded. These categories are known all the way up through the hierarchy to the root manager, forming a rather unspecified hierarchical inventory of services. The complete description of agent services is stored locally in the particular agent (service provider). This set of information can be accessed with a navigator tool to present it nicely to the user. So, agents can be asked what kind of services they provide or, more general, the service categories of a marketplace can be request. Since a service can be one of two possible types - commercial services and free services, this information will be brought to the user as well.

1.3.7 Routing of messages

There are a number of ways agents can communicate with each other, first of all they can speak directly, they can use a third party e.g. a manager agent or they can use a communication method like a black board. In JIAC, the agents use either one of first two methods.

Let us first consider how agents can be found on a marketplace.

An agent address consists of the attributes *host name*, *marketplace name* and an *agent name*. So, if all agents know the full addresses of all agents they like to communicate with, the agents simply open a communication channel to the agent in question. If that is not this case, the agents have to route their messages. Since all agents know at least their superior agent (usually the manager agent), they use it as a kind of relay station.

The routing process which is initiated by the manager agent takes advantage of the hierarchy of marketplaces and the fact that, the manager agent knows all agents on its marketplace as well as the manager agents above and below it. So, the routing takes place as follows: the first step is sending the message to the manager agent, then the manager is routing it to the appropriate manager of the destination agent, which will send the message to the receiver.

1.3.8 Components of Agents

A JIAC agent exists as a combination of several *components*. These components are exchangeable by the designer and can additionally be exchanged the agent itself. This means an agent is able to change its 'equipment' (i.e. communication or security component) even during runtime.

Each JIAC component can be chosen from an open set of components and encapsulates a certain semantical aspect of the agent. The components are plugged into a message bus that arranges the distribution of the internal message traffic of the agent. So the message bus forms the most basic part of an agent, its backbone.

This component based agent architecture brings all of the known advantages of component based software design, as there are higher reuse of software, more reliable and secure systems as well as faster realization of the software.

Each component has basic functionality for the support of:

- **Persistence** and **migration** of the agent. This includes also the availability of scripts, which represent the capabilities or services agent in JIAC has.
- **Interface for Integration** into the agent. This means implementing and extending the Java beans concept. **Concurrency**, which is realized by the fact that each component is a concurrent thread inside the agent.

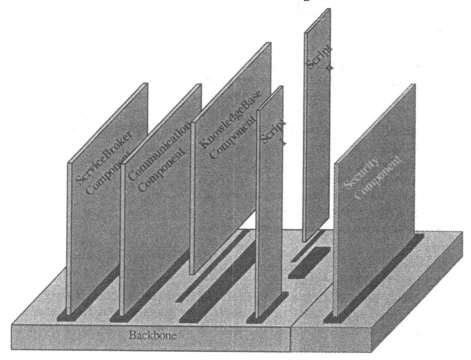

Fig. 4. Components of an agent

Reflecting the several aspects of an agent there are a set of possible components for agents in JIAC:

- A **ServiceBroker** looks up for a certain **service** given a certain speech act. The servicebroker can be viewed as a kind of name server for agents offering services.
- A **Communication Component** represents the communication facility to other agents and includes features like different communication protocols.
- An **I/O Component** encapsulates the complete access to other systems such as databases, legacy systems (e.g. HTTP server access) etc.
- A **Planner Component** encapsulates the abilities of the agent to act autonomously.
- A **Billing Component** takes care of the mapping of the services and actions taken place as recorded by the accounting component to a bill expressed in a certain currency.
- An **Accounting Component** logs all actions that the agent makes and all services it consumes.
- A **Debugging Component** represents the debugging facilities of the agent. Debugging is one of the main aspects of JIAC to encourage secure, reliable applications.
- A **Knowledge Base** contains the facts of the environment the agent acts in. That covers dynamic as well as static knowledge.
- A **Runtime Component** that controls the execution of scripts.

1.3.9 Communication Protocols

Agents under JIAC have to use given resources like Inter- or Intranet. To migrate from one host to another the agent has to be able to communicate via several protocol formats. Since JIAC is a Java architecture, there is a fairly big variety of different protocols JIAC can handle. Right now JIAC offers the following protocols:

- CORBAs IIOP
- TCP/IP
- KQML for speech acts
- Java's RMI (Remote Message Invocation)
- SSL (Secure Socket Layer)
- A Java-Beans conform component model.

1.3.10 The Tools of JIAC

As already mentioned above, JIAC is not restricted to some core classes, but it aims to include also a set of tools to support the agent developer in the process of design. As mentioned there are different phases of this process (agent-oriented analysis, agent-oriented design and finally agent-oriented programming) each tool focuses on different aspects within these phases.

Right now it is planned to equip JIAC with the following tools:

- An **agent oriented debugger**, that supports special features for agent oriented software structures. It is essential to give the agent developer a dedicated tool to find problems within the structures of the agent scenario, that are hard to find using conventional tools. A prototype is implemented right now that features basic functionality to debug agents.

- A **monitoring tool** that visualizes the agents within a scenario and shows their actions on the screen by animations. This tool offers functionality to erase mobile agents, to recall mobile agents from marketplaces and introspect agents located on a certain marketplace.
- A **constructor tool** is planned to support the agent programmer during the design and maintenance of the agents. Features like editing or exchanging the scripts of an agent, triggering the goal or planning behaviour of the agent will be convenient to handle by using a graphical interface.
- An **I/O tool** is planned to support the programmer in maintaining and equipping the agent with the necessary I/O facilities like accessing HTTP servers or certain protocols.

1.3.11 Applications of JIAC – Examples

The main area for JIAC applications is in the field of electronic commerce and telecommunication services. For these areas it is essential that the software is able to provide sophisticated functionalities to log the actions and to bill the used services.

Such applications can be:

- **Access tools** for networks and their services.
- To **support of mobile clients**, a personal information manager acts as a network resident personalized communication and information platform for mobile users.
- **Electronic commerce** on an agent oriented platform is facilitated by agents that offer and request services. Here it is very important to include existing environments like T-Online.
- **Service and network management** in the context of providing data services in intelligent networks but also generally for remote maintenance of network communication resources.

In JIAC, agents include an accounting component as well as a billing component. Those components can be exchanged independently and realize a customer specific accounting and billing environment without changing the core of the agent. The accounting component is responsible for keeping track of all the actions taken place by the agent and the used services. Furthermore, it is also bundling all the indirectly stimulated operations by other subscripts or agents that are needed to fulfil the specified goal of the agent. The billing component is now the part of the agent that is billing the customer for the consumed services. It is essential to separate these two tasks, since the accounting structure for one service may be the same, but the billing structure (i.e., the prices or costs) may be fundamentally different. To give an impression on how these components may act together in a real application, the following example is shown.

In the this example two marketplaces exist. One marketplace contains a number of service providing agents for example giving information about the departure time of certain planes and trains. On the second marketplaces is the user agent handling a graphical user interface for the users input. The user is able through the use of this agent to phrase questions about travel connections involving trains or planes. If a question is given to the user agent a mobile agent is created which will migrate to the first marketplace. On this market the agent will gather the information answering the users question from the service provider agents. When the agents finishes it will migrate back to the users market letting the user agent present the information. Also the user might get a bill from the service provider agents for the use of there services. JIAC will support building this example in the following ways. First JIAC provides mobile agents and stationary agents the programmer can choose from. He only has to write or, if possible, reuse existing scripts for these agents to realize the application. In particular the user interface, the strategy for the mobile agent for asking the providers and the provider scripts itself will be necessary to program. But no attention must be paid to the migration process, to the marketplaces and to the accounting process.

In real applications the different agents will surely not be programmed from one party alone but will, for the above example, come at least from the user interface programmers and from the service providers. So the idea is to let the service providers make agents offering there services and to have a number of application programmers to write a number of different applications using these services. Perhaps even other service providers may offer value added services based on the original services.

1.3.12 JIAC - a resume

JIAC is a reliable and safe environment for mobile agents under Java, that is especially suitable for applications concerned with electronic commerce and telecommunication services. For these purposes, JIAC comes with a complete set of classes for agents and marketplaces.

Whats more, JIAC is designed from the bottom up following a very modern component approach that makes it possible to build agents and marketplaces by reusing tested components which ends up in more reliable and more cost efficient systems. This plug-and-play approach makes JIAC a very promising candidate for further agent oriented software projects and future research. Since JIAC is an ongoing project with several applications, it is constantly extended by capabilities and functionality. For the next milestones we are working mainly in the field of the planning abilities of the JIAC agents and the implementation of graphical tools for agent oriented programming

Conclusions

In the introduction it was shown that agent oriented techniques are a good idea to address the demands of todays telecommunication market. Especially with the help of mobile agents and electronic marketplaces, as offered by JIAC, the telecommunication market is well suited to build those applications the user finds appropriate for the dawn of the 21st century.

2 References

[Brooks, R. A. (1986)], "A Robust Layered Control System for a Mobile Robot", IEEE Journal of Robotics and Automation 2 (1), 14-23.

[Brooks, R. A. (1991a)], "Elephants Don't Play Chess", In Maes, P. (ed) (1991), Designing Autonomous Agents: Theory and Practice from Biology to Engineering and Back, London: The MIT press, 3-15.

[Brooks, R. A. (1991b)], "Intelligence without Representation", Artificial Intelligence 47, 139-159.

[Brooks, R. A. (1991c)], "Intelligence without Reason", In Proceedings of the 12th International Joint Conference on Artificial Intelligence, Menlo Park, CA: Morgan Kaufmann, 569-595.

[Bond, A. H. & Gasser, L. (1988)], Readings in Distributed Artificial Intelligence, San Mateo, CA:Morgan Kaufmann.

[Bratman, M. E., Israel, D, J. & Pollack, M. E. (1988)], "Plans and Resource-Bounded Practical Reasoning", Computational Intelligence 4, 349-355.

[Carver, N. & Lesser, V. (1995)], "The DRESUN Testbed for Research in FA/C Distribution Situation Assessment: Extensions to the Model of External Evidence", In Proceedings of the 1st International Conference on Multi-Agent Systems (ICMAS-95), San Francisco, USA, June, 33-40.

[Carver, N., Cvetanovic, Z. & Lesser, V. (1991)], "Sophisticated Cooperation in Distributed Problem Solving", in Proceedings of the 9th National Conference on Artificial Intelligence 1, Anaheim, 191-198.

[Chaib-draa, B., Moulin, B., Mandiau, R. & Millot, P. (1992)], "Trends in Distributed Artificial Intelligence", Artificial Intelligence Review 6, 35-66.

[Chapman, D. (1992)], "Vision, Instruction and Action", London: MIT Press.

[Davis, R. & Smith, R. G. (1983)], "Negotiation as a Metaphor for Distributed Problem Solving", Artificial Intelligence 20, 63-109.

[Decker, K. S. (1995)], "Distributed Artificial Intelligence Testbeds", In O'Hare, G. & Jennings, N. (eds.), Foundations of Distributed Artificial Intelligence, Chapter 3, London: Wiley, forthcoming.

[Decker, K. S. & Lesser, V. R. (1993)], "Designing a Family of Coordination Algorithms", Proceedings of the 11th National Conference on Artificial Intelligence, Washington, 217-224.

[Dent, L., Boticario, J., McDermott, J., Mitchell, T. & Zabowski, D. A. (1992)], "A Personal Learning Apprentice", In Proceedings of the 10th National Conference on Artificial Intelligence, San Jose, California, AAAI Press, 96-103.

[Doran, J., Carvajal, H., Choo, Y. & Li, Y. (1990)], "The MCS Multi-agent Testbed: Developments and Experiments", in Deen, S. (ed.), Cooperating Knowledge based Systems, Heidelberg: Springer-Verlag, 240-251.

[DPE] Architecture – Engineering Modelling Concepts , P. Graubmann, W. Hwang, M. Kudela, et. al.
    TINA-consortium, document-label TB_NS.005_2.0_94

[Durfee, E. H. & Montogomery, T. A. (1989)], "MICE: A Flexible Testbed for Intelligent Coordination Experiments", In Proceedings of the 1989 Distributed Artificial Intelligence Workshop, 25-40.

[Durfee, E. H., Lesser, V. R. & Corkill, D. (1987)], "Coherent Cooperation among Communicating Problem Solvers", IEEE Transactions on Computers C-36(11), 1275-129
[Gasser, L. (1991)], "Social Conceptions of Knowledge and Action: DAI Foundations and Open Systems", Artificial Intelligence 47, 107-138.

[Gasser, L. & Huhns, M. (1989)], Distributed Artificial Intelligence 2, San Mateo, CA: Morgan Kaufmann.

[Gasser, L., Braganza, C. & Herman, N. (1987)], "MACE: A Flexible Testbed fo Distributed AI Research", In Huhns, M. (ed.), Distributed Artificial Intelligence, Research Notes in Artificial Intelligence, London: Pitman, Chapter 5, 119-152.

[Gasser, L. ,Rosenschein, J. S. & Ephrati, E. (1995)], "Introduction to Multi-Agent Systems", Tutorial A Presented at the 1st International Conference on Multi-Agent Systems, San Francisco, CA, June.

[Hewitt, C. (1977)], "Viewing Control Structures as Patterns of Passing Messages", Artificial Intelligence 8(3), 323-364.

[Huhns, M. N. & Singh, M. P. (1994)], "Distributed Artificial Intelligence for Information Systems", CKBS-94 Tutorial, June 15, University of Keele, UK.

[Jennings, N., Corera, J. M., Laresgoiti, L., Mamdani, E., Perriollat, F., Skarek, P. & Varga, L. (1995)], "Using ARCHON to Develop Real-World DAI Applications for Electricity Transportation and Particle Accelerator Control", IEEE Expert, Special Issue on Real World Applications of DAI systems.

[Lesser, V. & Corkill, D. (1981)], "Functionally Accurate, Cooperative Distributed Systems, IEEE Transactions on Systems, Man, and Cybernetics C-11(1), 81-96.

[Rao, A. S. & Georgeff, M. P. (1995)], "BDI Agents: From Theory to Practice", In Proceedings of the 1st International Conference on Multi-Agent Systems (ICMAS-95), San Francisco, USA, June, 312-319.

[Rosenschein, J. S. (1985)], "Rational Interaction: Cooperation Among Intelligent Agents", PhD Thesis, Stanford University.

[Rosenschein, J. S. & Zlotkin, G. (1994)], "Rules of Encounter: Designing Conventions for Automated Negotiation among Computers", Cambridge: MIT Press.

[Shoham, Y. (1993)], "Agent-Oriented Programming", Artificial Intelligence 60(1), 51-92.

[Smith, R. G. (1980)], "The Contract Net Protocol: High-Level Communication and Control in a Distributed Problem Solver", IEEE Transactions on Computers C-29 (12).

[Sycara, K. (1995)], "Intelligent Agents and the Information Revolution", UNICOM Seminar on Intelligent Agents and their Business Applications, 8-9 November, London, 143-159.

[Wittig, T. (1992) (ed.)], "ARCHON: An Architecture for Multi-Agent Systems", London: Ellis Horwood.

[Wooldridge, M. & Jennings, N. (1995a)], "Intelligent Agents: Theory and Practice", The Knowledge Engineering Review 10 (2), 115-152.

[Wooldridge, M. & Jennings, N. (eds.) (1995b)], "Intelligent Agents", Lecture Notes in Artificial Intelligence 890, Heidelberg: Springer Verlag.

[Wooldridge, M., Mueller, J. P. & Tambe, M. (1996)], "Intelligent Agents II", Lecture Notes in Artificial Intelligence 1037, Heidelberg: Springer Verlag.

[Zlotkin, G. & Rosenschein, J. S. (1989)], "Negotiation and Task Sharing among Autonomous Agents in Cooperative Domains", Proceedings of the 11th IJCAI, Detroit, Michigan, 912-917.

Author Index

Springer
and the
environment

 Springer

Lecture Notes in Artificial Intelligence (LNAI)

Lecture Notes in Computer Science